ECLIPSES

Astronomically and Astrologically Considered and Explained.

BY

SEPHARIAL

CONTENTS.

PREFACE.

THIS little work is designed to instruct the mind as yet unacquainted with the phenomena of eclipses, and to counteract the prejudice which affirms without examination that these phenomena are in no way connected with mundane events. In dealing with my subject I shall have occasion to speak of eclipses of the Sun and Moon not only in their physical causative relations, but also in their symbolical and prognostic relations.

If we can show that there are definite physical effects attending eclipses of the luminaries, it should be an easy matter to entertain the possibility that they may also have an effect in the psychic and mental spheres of our life, owing to the known relationship of mentality and organism.

That eclipses may have, further, a symbolical and prognostic value, follows from the fact that they can be calculated in advance, not only as regards the celestial position in which they occur, but also the mundane locality in which they will be witnessed, their duration and dimensions. For if we regard man as related upon all sides to his greater environment, represented by the celestial amphisphere, we may rationally suspect that changes taking place in that environment will produce in him a variety of corresponding changes and modifications.

And we have no reason to suppose that all eclipses will act upon him alike, seeing that they occur in different parts of the heavens and from different signs and constellations. Only from like causes we must argue like effects, and all that remains for us is to show that this is the case.

In this way it is possible to build up an *astrologia sana* as far removed from the superstition of the popular mind as most natural phenomena are from the folk-lore connected with them when they are properly studied and scientifically regarded.

The work here presented is intended to be an introduction to the proper study of eclipses rather than a complete exposition either in theory or practice. It is, in short, intended to show what eclipses are, how they are caused, the means of calculating them with adequate accuracy, and finally of finding a logical place for them in our scheme of thought as a part of that Cosmic Symbolism by which the Divine Intelligence is rendered to us intelligible.

London, 1915. SEPHARIAL.

ECLIPSES

ASTRONOMICALLY AND ASTROLOGICALLY CONSIDERED AND EXPLAINED.

CHAPTER I.

The Natural Causes of an Eclipse.

CONSIDER the Sun as the source of light, the Moon as a reflecting screen, and the Earth as an opaque body at one time in front of the screen, at another behind it. What are the phenomena we should expect from these relations?

(1) When the Sun's rays fall on one side of the screen and the Earth is on the other side of that screen, the result will be darkness, due to the fact that the interposing screen cuts off the rays of the Sun from the Earth. From the point of view of the Earth, the Sun is blotted out by the interposing body of the Moon. This is an eclipse of the Sun.

(2) When the Sun's rays fall upon the Earth, at such time as the Earth is between the Sun and Moon, the result will be that the image of the Earth will be thrown on the screen or body of the Moon, due to the fact that the Earth cuts off the rays of light which otherwise would be projected upon the screen. This is an eclipse of the Moon.

From the first phenomenon we see that the solid body of the Moon is interposed between the Sun and the Earth, so that we on the Earth cannot see the Sun; and from the second phenomenon we see that the shadow of the Earth is thrown by the Sun's rays upon the body of the Moon, so that we on the Earth see the projection of the Earth's shadow.

But inasmuch as all light travels in straight lines, it will be seen that in order to obtain these phenomena we have to presume that all three bodies are in the same plane, or approximately so, and that a line passing through the centre of any two of the three bodies will cut into the third or other of them.

This will perhaps be best understood if we remark that at New and Full Moon, every month, the three bodies are in the same meridian or vertical plane, but not necessarily at the same time in the same horizontal plane. An eclipse can only happen when all three bodies are in line in the same plane—that is, simultaneously in the same vertical and horizontal lines. For it is obvious that if two bodies lie in the same plane, a third body may be either above or below that plane, so that a ray of light or a straight line carried from the first of them and falling upon the second would not, if continued, also fall upon the third body.

Consequently, there is not an eclipse of the Sun every New Moon day, and there is not an eclipse of the Moon every Full Moon day. The reason for which is this:—

The Sun apparently moves in an orbit which may be regarded as an ellipse, of which the Earth occupies one of the foci. The plane of this orbit is inclined to our Equator at an angle of about 23°. But the Moon also appears to move about the Earth in an elliptical orbit, of which the Earth is also the focus, and the plane of this orbit is inclined to our Equator at an angle of about 28°, and, consequently, is inclined to the plane of the Sun's orbit at an angle of about 5°.

The point where the plane of the Sun's orbit and that of the Moon cross one another is called the Node. It is only when the Sun and Moon form a conjunction or opposition—that is, when there is either a New or Full Moon—on or close to the Node, that there can be an eclipse.

The New Moon when thus falling will constitute an eclipse of the Sun, and the Full Moon when thus falling will constitute an eclipse of the Moon.

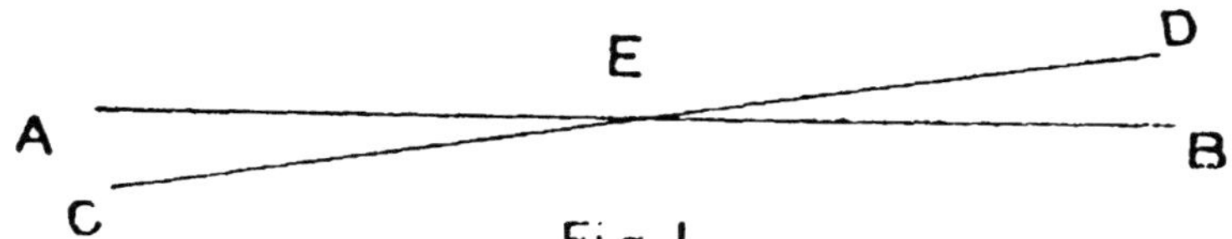

Fig I

If we take the plane of the Sun's orbit as the line A—B in Fig. 1, and the plane of the Moon's orbit as the line C—D, then the point of intersection at E will be the Node. Now if the Sun and Moon were only points in space, they could only coincide when both were simultaneously at the point E in their respective orbits. But they are bodies having apparent dimensions, and are approximately of the same apparent size. Therefore, if we suppose the Sun to proceed along the line E—B, and the Moon to proceed along the line E—D, it will depend on the apparent size of their discs and their combined motions as to when or at what point their discs will disengage so as to again render them two distinct discs. And this apparent meeting and separation of the two discs of the Sun and Moon will depend chiefly upon the motion of the Moon, since the Sun moves only 1° in its orbit in the course of a day, while the Moon moves about 13° in the same time, and hence it very soon leaves the Node and separates from the plane of the Ecliptic.

The distance between the plane of the Sun and that of the Moon is called the latitude of the Moon, and this is 0° at the Nodes and about $5^\circ\ 35'$ at 90° from the Node. The latitude will therefore depend on the distance of the Moon from the Node.

The North Node is the point at which the Moon crosses the Sun's path in its passage from South to North, and the South Node is that at which the Moon passes from the North to the South of the Sun's orbital plane. They are called respectively the Ascending and Descending Nodes of the Moon, and were anciently known as the Dragon's Head and the Dragon's Tail. These terms were derived from the ancient Chinese conception which regarded the course of the Moon as following the coils of a snake, the Head being originally in the constellation answering to the sign Scorpio (*Draconis*), and the Tail in the opposite point of the heavens. The symbol is an extremely luminous one. In Hindu astronomy these two points are known as Rahu and Ketu. In the Latin vorks they are referred to as *Caput Draconis* and *Cauda Draconis*.

It has been said that the Sun and Moon are apparently of about the same size. This means that at their respective distances from the Earth they appear to have the same amplitude in the heavens. And by this we see that the Sun is greater than the Moon in reality, being further from the Earth and yet appearing to be about the same size. But on close examination of the various solar eclipses it will be seen that sometimes the Moon covers the whole disc of the Sun and sometimes leaves a small outer ring of the Sun uncovered. The first case is called a total eclipse and the second an annular eclipse. Here the word "annular" means "like a ring," from the Latin word *annulus*. The reason for this difference in the apparent size of the Moon at different times may now be explained.

A body diminishes in apparent size in proportion to its distance from a spectator. If the Moon moved round the Earth in a circular orbit of which the Earth was the centre, it would always be of the same apparent size, but owing to the fact that from the point of view of the Earth it moves in an ellipse, in which the Earth occupies one of the foci, the Moon appears largest when in its perigree and smallest when in its apogee. The perigee is that point in its orbit where the Moon is nearest the Earth, and the apogee is always the opposite point, where it is furthest from the Earth.

Similarly, the Sun, from the same point of view, appears to move in an elliptical orbit about the Earth, and is consequently always altering its apparent size.

It will therefore depend on the positions of the Sun and Moon in their orbits at the times of their conjunctions and oppositions, as seen from the Earth, as to what their relative apparent sizes may be. It is customary to measure these bodies by their apparent semidiameters—that is, half the distance of their greatest width. This measurement is made in minutes and seconds of space. The mean semidiameter of the Sun is 16′ 6″, and that of the Moon 15′ 41″. The least of the Sun is 15′ 50″ and of the Moon 14′ 54″, while the greatest of the Sun is 16′ 23″ and of the Moon 16′ 39″.

If therefore the Moon and Sun should both be in

their apogee, that is, furthest from the Earth, at the time of their conjunction, the Moon's disc would not be large enough to cover that of the Sun, and the result would be a ring or annular eclipse. But if both the luminaries should be in their perigree, nearest the Earth, at the time of their conjunction, the disc of the Moon would be more than sufficient to cover that of the Sun, and the result would be a total eclipse.

We are, of course, speaking of ecliptic conjunctions only. There is a mensual conjunction of the Sun and Moon (New Moon), in which the two bodies are in the same longitude as seen from the Earth, but it is only when they are at the same time in the same plane, that is, when the Moon has no latitude, that an ecliptic conjunction is formed.

This can only happen at definite times, owing to the inequality of the relative motions of the Sun, Moon, and Earth. The ancients very early discovered this fact, and as far back as the year 2154 B.C. we find an eclipse of the Sun figuring in the Chinese history. It was a criminal offence even at that remote date for an astronomer to neglect to notify the approach of an eclipse, and consequently the Observer and Recorder, having failed to notify this eclipse, were both banished from the kingdom. Owing to the faulty condition of our Tables, great doubt has been thrown on this and other eclipses recorded by the ancients, and our astronomers have been forced to the ridiculous conclusion that because their Tables of the Sun and Moon motions were correct for observations made during the present century, they must necessarily be correct for all past time. This my New Astronomical Tables show to be a false conclusion, and they not only prove that this eclipse took place as recorded, but also other eclipses contained in various classical writings were either gross inventions or were actual observations. To argue that they did not take place at a time of day or night when they could have been observed in definite localities is to infer that they were retrogressively calculated, and this, so far as the Chinese eclipse is concerned, is to credit the ancient astronomers with a knowledge of celestial motions equal to that which has only recently been obtained by direct observations made with the most perfect instruments. However, to return to our Ecliptic Period.

It has been found that the Moon's Node moves backwards in the signs at the rate of $19^\circ 20'$ per year, and that the Moon and Sun are conjoined in the same part of the zodiac every 19 years. The Moon takes 29½ days to complete the course from one conjunction with the Sun to the next; this is called a Lunation. But during this time the Node will have retrograded, while the Sun will have gone forward in the zodiac. The period required for the Moon to pass from the Node to the same Node again will therefore be less than its complete revolution. The period is in fact about 27 days and one-tenth. Also it is seen that the Sun passes from the Moon's Node round to the same Node again in 346 3-5 days. These three periods, expressed in decimals of a day, are: The Draconic month 27.2 days, the lunation 29.5 days, the Sun's conjunction with the Node 346.6 days. Calculation will show that 242 Draconic months, 223 lunations, and 19 returns of the Sun to the same Node are all completed in the same time, or very nearly so. The period is 6,585 and one-third days roughly. This is equal to 18 years, 10 days, and about 8h. 9m. 35s. Consequently, if we add this period to the date of any eclipse we shall find another at about the same time. It cannot be exactly the same time, owing to the fact that the Sun is not in the same part of its apparent orbit at each successive return to the same Node, and the relative motions of the Sun and Moon are therefore altered. But it is near enough to have been observed useful for general purposes, and, as stated, was well known to the ancients. Now supposing that an eclipse of the Sun took place at sunrise, another eclipse would take place after the lapse of 18 years 10 days near noon or in the early afternoon, and thus would be visible in the same locality. But the next eclipse after a further lapse of 18 years 10 days would happen after sunset at that place, and consequently would not be visible there. But as the increment of 8h. 10m. over and above the period of 18 years 10 days is about one-third of a day, it will follow that after 54 years 1 month (18 years 10 days multiplied by 3) the eclipse will again come into the field of visibility. But at each such return there will be a difference in the dimension of the eclipse, owing to the fact that the Sun is getting further away from the Node. There can be no eclipse of the Sun when the latter is more than 18° 36′ from the Node. Thus there was a solar eclipse on July 8th, 1842, and after 54 years 1 month we find another on August 9th, 1896. In 1850, on August 7th, there was an eclipse of the Sun, and 54 years 1 month later there was another, September 9th, 1904. If all the factors were constant and their recurrence exactly synchronous, it would follow that at the end of 649 years (12 times 54 years 1 month) the eclipses would recur in the same order. In the course of subsequent chapters it will be shown why eclipses of the Sun and Moon cannot recur indefinitely in the same sequential order.

CHAPTER II.

Eclipses of the Sun.

FOR several reasons eclipses of the Sun have claimed more study of recent years than those of the Moon. One of these reasons is that they afford better opportunities of testing the inequalities of the Moon's motion and of adjusting the several equations by which the Moon's true place is found. Another reason is that they have contributed very largely to our material in the newly-developed science of Solar Physics. In a solar eclipse we get the interposition of an actual body, while in a lunar eclipse we only get the effects of the Earth's shadow upon the Moon. And because the Moon is our nearest neighbour and very accessible to our best telescopes we know more of her than we do of the great centre of the system to which our world belongs.

Eclipses of the Sun may be total, annular, or partial. The total eclipse occurs when the Sun and Moon form their conjunction within 5° of the Node, and at a time when the apparent diameter of the Moon is greater than that of the Sun. An annular eclipse likewise requires that the Sun and Moon shall be near the Node, but the apparent diameter of the Moon must be less than that of the Sun. A partial eclipse takes place when the Sun is conjoined with the Moon at more than 5° from the Node.

An eclipse that is total at one place on the Earth may be partial only at other places. Owing to the

Moon's motion in the zodiac being about 13° a day, and that of the Sun about 1° only, the Moon will overtake the Sun at the rate of about 12° a day, or 30′ in an hour. The Earth is all the time revolving on its axis from West to East, and the luminaries will be carried by apparent motion from East to West. But at the point of time when the centre of the Moon's disc coincides with the centre of the Sun's disc, the conjoined luminaries will be exactly overhead at some place on the Earth's surface, and at this place the eclipse will be "central at noon." But it will be seen that, as the combined apparent diameters of the Sun and Moon amount to about 1°, an eclipse of the Sun will begin when the Moon is about half that distance, or about 30′ to the West of the Sun, for then their cusps will come into touch. In the course of half an hour the Moon's limb or cusp will have touched the centre of the Sun's disc, and in another half-hour the centre of the Moon will coincide with the centre of the Sun. This phase will only last for a minute or two, and the Moon will then begin to move off to the East of the Sun, taking about the same time to complete the other half of its transit. Thus the eclipse would take two hours from first to last contact if the Sun remained still and the Moon passed exactly over its centre. But the Sun is moving by apparent motion in the zodiac all the time, and in the course of two hours will have increased its longitude by some 5m. of space, and the Moon will require another 10m. of time to overtake this. Moreover, the Moon is not always progressing at the same rate, being at the time of eclipse at a certain place in its orbit in one instance and in another part of its orbit on another occasion. Hence we have to find, first of all, the apparent semi-diameters of the Sun and Moon and add them together, and then find the rate at which the Moon overtakes the Sun in longitude. We shall then get the time from the first contact to the centre of eclipse, which will be half its complete duration.

Now these elements can be extracted from the Ephemeris for the year, and they can then be referred to the table which is appended to this work.

There was an eclipse of the Sun on April 17th, 1912. This we know from the fact that it was a New Moon on that date and that the Sun was then (at noon) in Aries 27° 6′, while the Node was in Aries 21° 25′. The almanac tells us that the New Moon or conjunction in longitude was at 11.40 a.m.

At this time the Sun and Moon were in Aries 27° 5′. The Moon had 0° 31′ North latitude. This corresponds with the 6° that the luminaries are distant from the Node at this time. (See Table, p. 42, *Ready Reckoner*.) Now if we were looking at the conjoined luminaries from the plane of the ecliptic, they would be in conjunction, for they would occupy the same celestial meridian or longitude. But as we are regarding them from the plane of geographical latitude, the Moon will not be on the same geographical meridian as the Sun at 11.40 a.m., but a little to the West of it (as shown below), and we therefore have to wait for the Moon to increase its longitude sufficiently to bring it into the same geographical meridian as the Sun. This is found by the coincidence of the Right Ascension. Thus at 11.40 a.m. the Right Ascension of the Sun was 25° 8′, and that of the Moon was only 24° 57′. The difference is 11′, and as it would take the Moon 22m. of time to increase its Right Ascension by this amount, we have to add 22m. to 11.40 a.m., and thus get 12.2 p.m. as the time of their conjunction in Right Ascension, when they will be on the same geographical meridian at the same time. At 12.2 p.m. the Moon's longitude was Aries 27° 17′, its latitude N. 0° 32′, and its declination 11° 0′ N. Its Right Ascension was therefore 25° 8′, which is the same as that of the Sun.

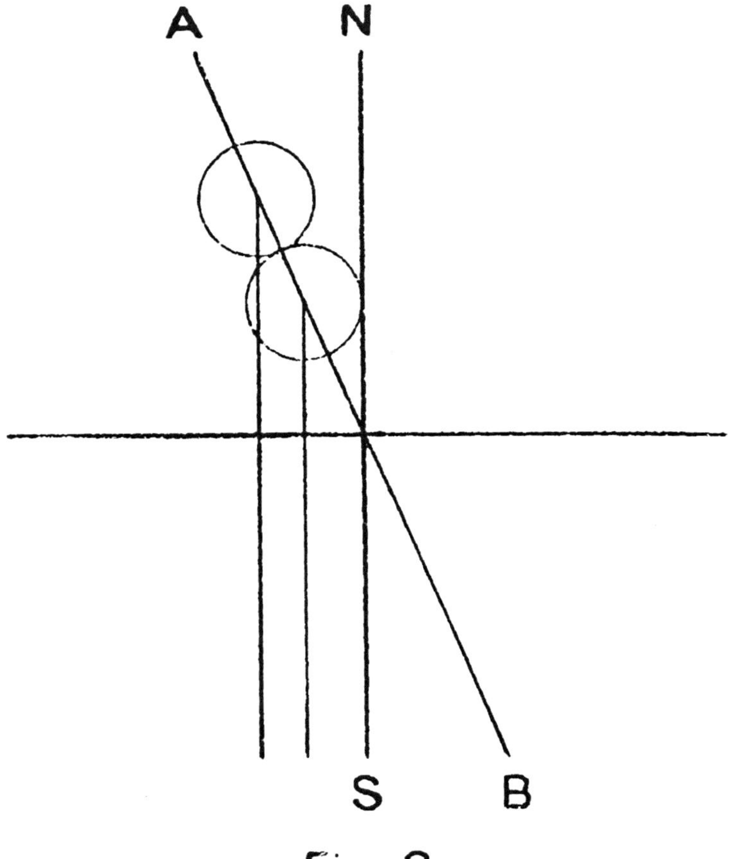

Fig 2

Then it will be seen that as the Moon has 11° of N. declination, while the Sun at the same time has only 10° 27′, their bodies, although having their centres on the same meridian, will not coincide with the same geographical latitude. Hence, although the Sun is at the zenith, or immediately overhead in latitude 10° 27′ N. (answering to its declination), the Moon will not be overhead there, but at a place that is 33′ further North. Then if we draw a line from the centre of the Sun through the Moon to the Earth, we shall strike a place in North latitude that will observe the discs of the Sun and Moon at this time to exactly coincide, and at that place the eclipse will be "central at noon." This is very easily found by dividing the sine of the difference of the declinations of the Sun and Moon by the sine of the Moon's horizontal parallax, and the result will be the sine of the angle formed by the line passing from the Sun through the Moon to the Earth, which, being applied to the Sun's declination, will give the geographical latitude at which the eclipse is central.

The horizontal parallax of the Moon is determined by its place in its orbit. This is counted from the apogee, or furthest distance from the Earth, and is called the Moon's Anomaly. This will be found, together with the corresponding horizontal parallax, in the Table, page 44, of my *Ready Reckoner*. This is found to be 58′ horizontal parallax. Then we have :—

Diff. of Declination 33′	sine log.	7.98223
Hor. Par. of Moon 58′	sine log.	8.22713
Angle of Apparation ...	sine log.	9.75510 = 34° 41′

and as the Sun has N. declination	10° 27′
Add	34 41
Geo. Lat. North	45° 8′

The Right Ascension of the Sun was found to be 25° 8′, which, converted into Sidereal Time is 1h. 40m. 32s., and the Right Ascension at noon was

1h. 40m. 59s., and at two minutes past noon 1h. 42m. 59s., and the difference between this and the Sun's Right Ascension is ·2m. 27s., which gives about $\frac{1}{2}^{\circ}$ of longitude West for the meridian of conjunction in Right Ascension. We therefore get a geographical position in the South-West of France as the place of centrality at noon. The values being approximate, the position is also approximate.

Now to find the duration of the eclipse we have to make a calculation of the anomalies of the Sun and Moon, *i.e.*, their distances from the apogee, and hence their places in the orbit. From the Tables at the end of this work it will be found that the Sun's anomaly is 9s. $15^{\circ}\ 41' = a$, and the Moon's is 3s. $8^{\circ}\ 27' = A$. Also it will be seen from the Table of Eclipse Elements that the semidiameter of the Sun is $16'\ 1''$, and that of the Moon is $15'\ 48''$. From this we know that the Sun appeared larger than the Moon, and consequently the eclipse was not total even where central at noon, but annular. For this reason it was an extremely interesting phenomenon to astronomers, and students of solar physics made special trips to the North of Spain on this occasion.

Now by adding the semidiameters of the Sun and Moon together we find they amount to $31'\ 49''$, which is the distance the Moon has to travel from first contact to central eclipse, and this will be half the duration of the phenomenon. To find this time we again refer to our Tables under the columns of "hourly motion," and against the respective anomalies of the Sun and Moon we find for the Sun $2'\ 27''$ and for the Moon $33'\ 55''$, which being compared show a difference of $31'\ 28''$, at which rate the Moon overtakes the Sun per hour.

Then, as we have the distance of first contact from centrality $=31' \ 49''$, we say by proportion :—

As $31' \ 28''$ is to 60m., so is $31' \ 49''$ to the number of minutes from central eclipse (12.2 p.m.) to the contact. This works out to 60m. 36s. We therefore know that the first contact was at 11.1 a.m., central phase at 12.2 p.m., and the last contact at 1.3 p.m., the whole duration being 2h. 1m. 18s.

We have now gone through all the elements which enter into the calculation of an eclipse of the Sun. This phenomenon has been under survey in all parts of the world from the very earliest ages of civilisation, and, as we have seen, was deemed of so great a consequence to religious and civic life in ancient China as to make it a capital offence to disregard the celestial portents.

The Chinese, indeed, have given us the longest and best authenticated list of eclipses of the Sun since the days of Wu Wang, B.C. 1122, and to these Confucius has also added a record of eclipses which took place during the period from 722 B.C. to 494 B.C. as derived from the archives of the Princes of Lu, his native province. This great scholar has also given us the historical record of the most ancient eclipse of which there is any mention. It is that of a solar eclipse in the reign of Chung Kang, B.C. 2154, to which reference has already been made. For particulars, see my *Ready Reckoner* under the head of "Eclipses."

The next most ancient is that which took place in the reign of Yew Wang on September 5th at about 11.35 p.m., Greenwich time, in the year 775 B.C. This eclipse is referred to in the collection of poems known as the Shi King, preserved to us by Confucius.

The next most ancient is the Nineveh eclipse of

762 B.C., which happened in the reign of Uzziah. The eclipse took place on June 15th.

It should here be observed that in giving dates for astronomical purposes for years before our era, the secular date must be reduced by one year. This is necessary because astronomical values are taken from an epoch which is B.C.—o—A.D. Thus the year before A.D. 1 is, according to common chronology, B.C. 1, whereas it is astronomically A.D. o. In the case of the Nineveh eclipse the common chronology or secular date is given as B.C. 763, but this is astronomical B.C. 762, and the same with others already referred to.

The next eclipse of prominent record is the first of those recorded by Confucius, and which took place on February 22, B.C. 719. It links up directly with the famous Babylonian eclipses of the Moon which preceded the demolition of that Empire, for it was the solar eclipse immediately preceding the second of the lunar eclipses visible at Babylon, and which took place on March 8th, B.C. 780.

We may now pass on to the consideration of lunar eclipses, which afford considerable interest to astronomical students, and are as full of problems for the observer as are those of the Sun.

CHAPTER III.

Lunar Eclipses.

WHEN we come to the consideration of lunar eclipses, we are confronted by an entirely different set of phenomena to those which characterise eclipses of the Sun. It has already been shown that the essential difference of the two lies in the fact that in the solar eclipse we have the occultation of a luminary by a solid body, while in the lunar eclipse we have merely the shadow of a solid body thrown upon the surface of a reflecting screen.

We know the Sun to be the body from which light proceeds, and we also know that the Earth, as a satellite of the Sun, is very small in comparison. This gives rise to the fact that a cone of shadow is thrown by the Earth into space on the side furthest from the Sun. This is called the "dark cone." But as light travels in straight lines, there will also be another shadow in the shape of a cylinder, which exactly defines the dimensions of the Earth. These shadows are thrown out into space, and it is the passage of the Moon through them that constitutes a lunar eclipse. The shadows are always there, but the Moon is not always coincident with the plane in which they fall—that is to say, the plane of the ecliptic. For several successive revolutions the Moon will pass around the Earth, coming to the Full at points that are successively nearer to one of the Nodes, until at last a Full Moon is formed which is within 12° 24′ of the Node, and at this point the Moon touches the shadow of the Earth, or as it is said, enters the dark cone. The nearer to the Node the Full Moon is formed the greater will be the eclipse of the Moon.

In the following diagram (Fig. 3) let A—B be the plane of the ecliptic or Sun's annual path. Let E be the Earth and M the Moon at the Full. Then the orbit of the Moon about the Earth will be represented by m—n, and it will be seen that there are two distinct phases of the eclipse. The first is the penumbral eclipse, when the Moon enters the track

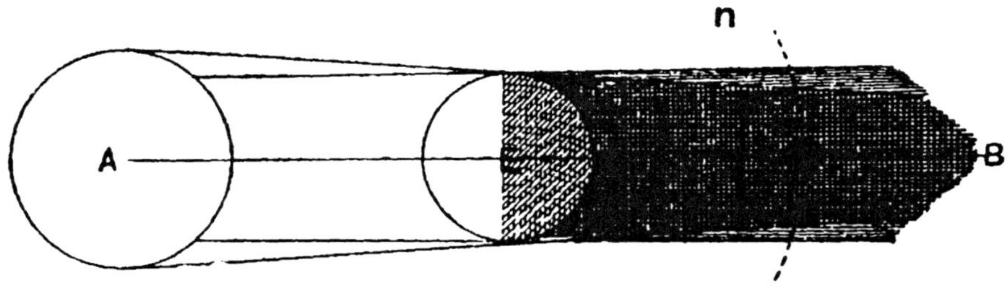

Fig 3

of the penumbra or light shadow, and the second is when it enters the cone or dark shadow. It may thus happen that the Moon may enter the penumbra and not the dark cone, as when it is more than 12° and less than 17° from the Node at the time of the Full. In such case there is only a blurring or obscuration of the Moon's brightness. But when within the prescribed limits of 12° from the Node, the Moon will show the clearly-defined cusp of the Earth's shadow upon its surface.

The duration of the eclipse will depend on two factors—namely, the size of the cone or shadow of the Earth, and the velocity of the Moon at the time of eclipse.

From what has already been said, it will be seen that it depends on the place of the Earth in its orbit and its consequent distance from the Sun as to what dimensions the Earth's shadow will have. In winter the shadow is largest, because the Earth is then nearer to the Sun. At the summer solstice the shadow will be least, because the Earth is then furthest from the Sun, and at the equinoxes it will be at the mean. The controlling factor is therefore the Sun's anomaly, which determines its place in the orbit and therefore its distance from the Earth. We are, of course, speaking of the Sun as if it were a moving body, because it is apparently so, and we are dealing with appearances when studying eclipses.

But also the Moon at the time of eclipse will be in a certain part of its own orbit, and therefore its motion will be greater or less than the mean. At the perigee its motion is greatest, at the apogee it is least, while at the distance of three signs from either point it is at the mean. The mean motion is 13° 10′ per day. The Moon's anomaly, therefore, will determine its velocity.

We therefore have to find the anomalies of the Sun and Moon at the time of the opposition, or Full Moon.

The Moon's anomaly will give us the semidiameter of the Moon and its horizontal parallax, or the semidiameter of the Earth's disc as seen from the Moon. It also gives us the Moon's hourly motion. From the Sun's anomaly we find its semidiameter and its hourly motion. From the Moon's distance from the Node we find its latitude. From the sum of the semidiameters of the Sun and Moon we obtain the semidiameter of the penumbra or light shadow. These may be found in the following order and tabulated for use as required :—

1. The true time of Full Moon.

2. The Moon's anomaly, its horizontal parallax and semidiameter; also its hourly motion.

3. The Moon's distance from the Node, and hence its latitude.

4. The Sun's anomaly, its semidiameter and hourly motion.

5. The semidiameter of the Earth's shadow at the Moon's orbit.

The time of Full Moon can be found from the almanac, or may be calculated from the "New Astronomical Tables." The Moon's anomaly for this time is calculated from the Tables at the end of this work, or from those in the "Ready Reckoner." The Moon's horizontal parallax, its semidiameter and hourly motion, all depending on the anomaly, will be found in the same Table. The place of the Node is calculated from the Node Table in the "Ready Reckoner." The latitude is found from the Table on p. 42 of the same work. The Sun's anomaly is calculated from the Tables annexed hereto, and this gives its semidiameter and hourly motion.

The semidiameter of the penumbra is found by adding together the semidiameters of the Sun and Moon.

The angle of the Moon's path in relation to the ecliptic is always taken at 5° 35′.

In every case the Moon must be less than 12° 24′ from the Node, in order that there may be an eclipse of the Moon at the Full. In this connection the Tables of lunations in connection with that of the Node given in the "Ready Reckoner" are very useful in the first instance in finding whether there will be an eclipse at any Full Moon.

Now let us take an example of an eclipse of the Moon (see Fig. 4), and apply these elements to the phenomenon. We shall then see how they are to be employed, and what bearing they have upon the matter.

There was a considerable eclipse of the Moon on September 4th, 1914. The Full Moon was at

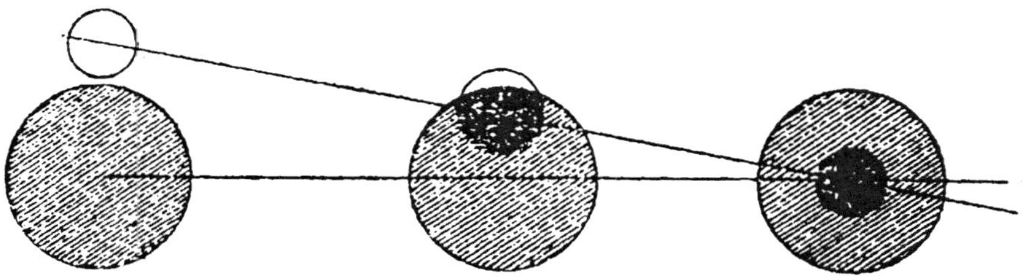

Fig 4

2.1 p.m. and the opposition in Right Ascension at 2.28 p.m. We may use the former time as being more accessible. The time of Full Moon and also that of the opposition in Right Ascension are given in the Nautical Almanac. But only the first of these is usually given in the popular almanacs.

Calculation shows the Sun at the given time to be in Virgo 11° 11′, and the Moon was therefore in Pisces 11° 11′. Calculation of the Moon's anomaly shows it to be 10s. 18° 36′, its Hor. Par. 55′ 16″, its semidiameter 15′ 4″, and its Horary motion 30′ 54″.

The Sun's anomaly is found to be 2s. 1° 25′, its parallax 9″, its semidiameter 15′ 58″, and its Hor. motion 2′ 25″.

Then to find the semidiameter of that part of the Earth's shadow through which the Moon passes :—

Add Sun's Hor. Parallax (always)	0′	9″
To Moon's ditto (here)	55	16
Sum	55	25
Subtract Sun's semidiameter ...	15	58
	39	27

The Moon therefore has to pass through a section of the cone which is 1° 18′ 54″ in diameter, the half of which will be the extent from the first contact to the middle of eclipse, and the other half from the middle to the last contact. We have therefore to find the time in which the Moon traverses this arc of 39′ 27″.

The Moon's Horary Motion was found to be	30′	54″
That of the Sun	2	25
Rate per hour at which Moon overtakes Sun	28	29

Then by proportion : As 28′ 29″ : 60m. : : 39′ 27″ to 82m. 48s. This we may call 83m. or 1h. 23m., and, being taken from the time of centrality, gives 12.38 p.m. (Greenwich mean time) for the first point of immersion in the shadow, 2.1 p.m. as the mean time of the maximum phase, and 3.24 p.m. as that of the final emergence. The eclipse would therefore last for 2h. 46m. roughly, and as the Sun was above the horizon during this period, the Moon must have been in the opposite point, and would not therefore be visible at Greenwich.

One other point has to be observed. It is that if the Moon's latitude is N. the south limb of the Moon will be immersed, while if the latitude is S. the north limb of the Moon will suffer eclipse. The Moon's longitude being Pisces 11° 11′, and that of the Node 5° $21'$ of the same sign, the Moon will have passed the Ascending Node, and will therefore be in North latitude, and its distance from the Node being 11° $11' - 5^\circ$ $21' = 5^\circ$ $50'$, that latitude will be 0° $30'$ $28''$, and this informs us that the Moon was eclipsed on the south limb.

For the dimensions of the eclipse we say :—

As the Moon's distance from Node 5° $50'$ is to the limit of totality (always) 5° $0'$, so is the unit of magnitude $= 1$ to 0.86, and by multiplying this by 12 we get rather more than 10 digits, which is the old method of computing the dimensions of an eclipse, the diameter being equal to 12.

There are many other points connected with the calculation of eclipses which are of interest, but which do not serve the main purpose of this work. One of these is the extent of territory over which any eclipse is visible. Without going into the figures required for this purpose, it will suffice if we here give some approximations which can be readily estimated.

In the case of the solar eclipse we have the hour-angle of the Sun, or its distance from the Greenwich meridian at the time of conjunction, to guide us in the estimate of the longitude at which the luminaries will be conjoined in mid-heaven. We have also the difference of the declinations of the Sun and Moon at the time, and the consequent angle of apparition formed thereby, to guide us as to the approximate latitude at which the eclipse will be central when on the meridian.

In the case of the lunar eclipse we have the hour-angle of the Moon to give us the longitude, and the declination of the Moon will show where it is over-head at the time of its meridian transit.

For the ulterior purposes of our present treatise

it is necessary to compute the magnitude of the eclipse and the decanate, or one-third section of the sign, in which the eclipsed luminary is found at the time.

So far as the geographical extent of the eclipse is concerned, it will be observed that during the time from the first contact, through the entire phase until the last contact, whether it be a lunar or solar eclipse that we are considering, the Earth will be turning on its axis all the while from West to East, and the successive phases of the eclipse will therefore be observed in a tract of country running from East to West.

But also it will be observed that in the case of a solar eclipse the more oblique the angle at which an eclipse is visible the greater will be the tract of country over which it will be seen. For at the zenith there would only be a shadow thrown over a limited area, while at the rising or setting the shadows would be longer, as we observe is the case when trees and other objects obstruct the sun's rays when it is near the horizon. And if this happened soon after sunrise, or when the Sun is in the Eastern quarter of the heavens, the shadow would gradually shorten and sweep Eastward as the Sun came towards the meridian. But the reverse of this would be the case if the Sun were eclipsed in the Western quarter of the heavens.

If the Moon kept always in the same plane during the course of an eclipse, and its relations with the sun were unaltered, the shadow would pass over a definite parallel of latitude. But the Moon is crossing the ecliptic at an angle of about $5^\circ\ 35'$, and hence it follows that the shadow takes a curve which is due to the combined motions of the Earth on its axis and the Moon in its orbit. So by taking out the elements for the beginning, middle, and end of an eclipse, and finding the angle of apparition for these three points of time, they will direct us to three localities, and an arc including them will define the approximate course of the shadow.

The point which we have to keep clearly in mind in the two kinds of eclipses, solar and lunar, is that in the solar eclipse the Moon's shadow is thrown on the Earth, while in the lunar eclipse the Earth's shadow is thrown on the Moon. In both cases the Sun is the source of light, and the abscission of its rays by an interposing body is the cause of the phenomenon.

CHAPTER IV.

Historical Eclipses.

MENTION has already been made of one or two eclipses which extend the records of astronomers back into the most remote times. All antediluvian records are of course lost, if there existed any such, but almost immediately afterwards we find Ta Yu, the Chinese patriarchal ruler, regulating the seasons by the stars and appointing observers and recorders of celestial phenomena as Officers of State.

This would be about the year B.C. 2205, when he began to reign over that part of China which is comprised between the Yang-tse-kiang and the Hwangho, the two great rivers of China. Considerable method and care appears to have been employed in this matter, for we find that in addition to the official Observer and Recorder there were four officers appointed over the four points of the compass to regulate the seasons and observe that the calendar at all times coincided with the transit of certain stars over the meridian at midnight. But it was to the official Recorder and Observer that the State looked for information as to the times when eclipses should occur.

In the historical classic " Shu King " we have the following interesting record of how two of these appointed officials failed in their duties through drunkenness, and how they were consequently expelled from the kingdom.

" At that time Hi and Ho corrupted their principles and abandoned themselves to wine. They neglected their offices, forsook their posts, began to confuse the celestial laws, and ignored their faculty. Consequently, when in the third month of autumn, on the first day of the month, there was a partial eclipse of the Sun in the constellation Fang, the blind musicians beat their drums, the lower officials went off on horseback, and the directors of the people ran afoot in general consternation. Hi and Ho were dead to their duties of office, not having known or heard of it, and were stupidly perplexed as to the celestial signs, so that they committed a crime which former rulers proscribed against. The regulations in regard to eclipses say : Being before the time, the astronomers are to be killed without respite, and being behind time, they are to be slain without reprieve."

Now here is a circumstantial account of a partial eclipse of the Sun seen in China in the reign of Chang Kang, for it was he who is reported to have uttered the above words, which modern astronomers have had the temerity to dispute, for the (to them) adequate reason that the eclipse cannot be found by their modern Tables. The constellation Fang is a group of stars in Scorpio which at the date of Chang Kang coincided with the equinox. Chang Kang began to reign in the year 2159 B.C. and extended his reign to the year 2146 B.C., when he was succeeded by Ti Siang. We have therefore to find a partial eclipse of the Sun which took place between the years 2159 and 2146 B.C. at the time of the equinox. The text says the "third month of the autumn." Now the Chinese began their year at the point when the Sun was half-way between the winter soltice and the spring equinox, which would be February 5th, and the New Moon of February was therefore the first day of the first month of spring, the lunation of May marked the first day of the first month of summer, that of August marked the first day of the first month of autumn, and consequently October was the "third month of autumn," and the lunation of that month is the one that we are concerned with. All we have to do is to find a lunation which took place in October close to the equinox between the years 2159 and 2146 B.C. There is only one such, and this was in October, 2154 B.C., close to the equinox, in the constellation Fang. The actual date was October 12th, and the time was between 7 and 8 in the morning at Pekin. Thus it was only three days from the autumnal equinox, which at that time fell on October 9th.

In the face of this consistent testimony, it is more than surprising that astronomical writers should have fixed upon an eclipse that took place in the year 2128 B.C., and later on one that occurred in 2136 B.C., neither of which are consistent with the historical record. The reason, however, is simple. Ferguson was not apparently aware of this eclipse, and possibly had no commerce with Chinese eclipses; at all events, he does not cite any in the course of his work, but his Tables show the eclipse clearly enough and the time, which is not quite accurate, is 7·34 a.m. at Pekin on October 12th, 2154 B.C. But the later tables, such as those published in the Encyclopædia in 1858, improved Ferguson's out of existence, so to speak, and incidentally showed this eclipse to have taken place "before sunrise at Pekin," and consequently it could not have been observed at that place. Then what are we to think of the historical record which was preserved to us by Confucius? We can only credit the Chinese astronomers with such great proficiency in the knowledge and use of laws of eclipses in times prior to Confucius that they were able to retrogressively determine that an eclipse had taken place in October, in the fifth year of the reign of Chung Kang, close to the equinox ! It is surely simpler, more direct, and more consistent with what we know of Chinese astronomy of that period, to suggest that the eclipse actually took place as recorded, that it was actually observed at Pekin, and that possibly our Tables are at fault. Now this appears to have been suspected, for we find that astronomers have introduced what is called a " secular equation " of the Moon's longitude and anomaly, from which it is obvious that they are convinced that formerly the Moon was moving quicker than it does at the present time, since the modern mean motions of the Moon are taken from actual observations, and therefore, according to Kepler's law, the Moon was nearer the Earth then than now, and moving in a smaller orbit. This fact I have brought out in my New Astronomical Tables, and incidentally have dispensed with about 37 equations of the Moon which hitherto have been required to compute the Moon's true longitude at eclipses in remote times.

The second most ancient eclipse on record is also due to the Chinese, and is mentioned in the " Shi King " as having taken place in the days of Yew Wang, who reigned between the years 781 B.C. and 770 B.C., being succeeded by the Emperor Ping Wang. This has wrongly been referred to the total eclipse of the Sun which took place about three-quarters of an hour after noon at Pekin on June 4th, 780 B.C. The Sun was a little more than 2° from the Node. Its semidiameter was less than 16′, and the Moon's was more than 16′, and as the Sun was then in Gemini 25 near its greatest declination North, the eclipse would be total and close to the zenith in Southern China. A further eclipse appears to have been the subject of a homily in the collection of poems contained in the "Shi King." It is there said that "In the tenth month at her conjunction on the first day called Sin-Mao, the Sun was eclipsed very badly, both the Sun and Moon being hidden, and this would be a cause of mourning among the lower classes. For the Sun and Moon being thus conjoined announce tribulation to the people if they do not amend their conduct. The four quarters of the kingdom are not regulated, and their good is not studied. An eclipse of the Moon is only a common thing, but this of the Sun is a portent of great evil."

This obviously does not refer to the eclipse of June 4th, 780 B.C., which Rev. S. J. Johnson has cited in his "Historical Eclipses" as that mentioned by Confucius which occurred in the reign of Yew Wang, for June is not the "tenth month," as stated in the text, and the day Sin-Mao did not then fall on June 4th. To make this intelligible to the reader it should be remarked that the Chinese had a sixty-year cycle, to each year of which a distinctive name was given, exactly as is done in India to-day, and also a sixty-day cycle, which ran successively through the year and came round again in twelve years to the same dates. Also it should be observed that whereas anciently the Chinese began their year with the first New Moon after February 5th, which was the first day of spring, at later times this beginning of the year was transferred to December, and was thus fixed in the days of Yew Wang. Therefore the "shih yue," or tenth Moon, referred to would fall on the New Moon of September. Now the day Sin-Mao fell on a New Moon day in September, 775 B.C., and there was an eclipse of the Sun, about half the disc being obscured. It is certainly not so important an eclipse as that of 780 B.C., but as it agrees with the data we are bound to accept it. Doubtless the writer of the poem had also witnessed the latter great eclipse of the Sun five years earlier, but lacked the inspiration of the moment to versify upon it, otherwise he would have referred to the conjunction of the seventh month, and not at all to the day Sin-Mao, which then fell 47 days earlier or 13 days later. But here we have Sin-Mao falling on the New Moon day of the tenth month, and there can be no shadow of doubt that this was the eclipse referred to in the Siao-Ya section of the "Shi King."

The next great eclipse on record is that of June 15th, 762 B.C., which occurred in the reign of the Assyrian king Assurdayan, who was contemporary with Uzziah, King of Judah. The record was discovered in the Assyrian Tablets in the British Museum, and was published in 1867 by Sir H. Rawlinson in *The Athenaeum*. The Tablet to which this refers had been already translated by Dr. Hinckes in 1854, but its historical setting had not been appreciated. The eclipse was eventually found to be of considerable importance in relation to Scripture history, for it occurred only 24 years after the prophecy of Amos. Sir H. Rawlinson says as follows: "In the month Sivan an eclipse of the Sun took place, and to mark the great importance of the event a line was drawn across the Tablet, although no interruption takes place in the official order of the Eponymes.

"Here then we have a notice of a solar eclipse which was visible at Nineveh which occurred within 90 days of the vernal equinox, and which we may presume to have been total from the prominence given to the record, and these are conditions which during a century before and after the era of Nabonassar (746 B.C.) are alone fulfilled by the eclipse which took place on June 15th, 763, (=762 astron.)." Now the eclipse was predicted at Samaria, and referred to Nineveh, at both of which places it would have been observed. This eclipse has, therefore, had the chronological importance of having been instrumental in determining the true dates of the kings of Israel, affording a correction of some 25 years.

The next most ancient eclipses are those given in the Syntaxis of Ptolemy, the astrologer and geographer, of Alexandria, who revived the records of the astronomer Hipparchus some three centuries before his time, and also made important astronomical observations of his own. From these we gather that, as Tycho, the Danish astronomer, remarked, "Hipparchus and Ptolemy would not have mentioned these eclipses (those of Babylon) had they been able to find records of earlier ones, since from them the motions of the Sun and Moon could have been more accurately determined." This gives the keynote to all astronomical research in connection with lunar motions connected with eclipse phenomena, and it cannot be said that the greatest use has been made of them, for instead of clearing the deck of all the encumbrances by which modern astronomers had sought to adjust their Tables to ancient observations, or their modern observations to ancient records, they have persisted in retaining the innumerable equations of Buckhardt, even going so far as to employ the action of the planets Jupiter and Venus upon the Moon, while leaving out of consideration entirely that of the other planets, and thus ignoring entirely the fact of equilibrium which underlies all the conclusions of Kepler and Newton. From false premisses they have only succeeded in erecting a superstructure of false equations, which break down as soon as they are applied to ancient eclipses. I affirm most distinctly that the relations of the Sun, Moon, and Earth are the only factors in the production of an eclipse, and that the Moon's motion in its orbit is controlled entirely by the joint action of the Earth and the Sun, and by no other body of the solar system, and that all the equations of the Moon's motion are due entirely to the action of the Sun, including its acceleration in its orbit known as the secular equation, and that these amount in all to five equations, which are constant and unvarying. By this concept of the true motion of the Moon, all the ancient eclipses fall out exactly as they are recorded, not only as to date and time, but also as to dimensions and place of apparition, and the same holds good in regard to modern or future eclipses, whether of the Sun or Moon. Nature is simple, not complex. It is our minds that are complex and tortuous, and this accounts for the scientific pains that have been taken to render astronomical calculations as complicated and intricate as possible. A mere schoolboy could make them with due precision and accuracy when properly instructed in the facts.

In the Syntaxis we have the record of Hipparchus and many current observations of Ptolemy, in all amounting to 19 entries. Among these there are two ancient eclipses of the Moon mentioned by Hipparchus. The first was observed at Babylon, 383 B.C., December 23rd, when the Moon "began to be eclipsed half an hour before the Sun rose." But by most of the Tables it will be found that the Moon on that date had set at Babylon before the eclipse began, so that there would have been no chance of observing it there. The second instance is one at Alexandria, 201 B.C., September 22nd, when the Moon " rose so much eclipsed that it must have begun half an hour before she rose "; but by the Tables it appears the eclipse did not begin until some little time after the Moon had risen at Alexandria. These valuable records set astronomers thinking, and there now remains no doubt whatsoever that the Moon's acceleration has diminished since those days, and that the mean value of a lunation, as now observed, has to be diminished for ancient times. The diminution is about 4m. per century. The exact values are given in my New Astronomical Tables, to which the student is referred. They are incorrectly stated in modern Tables owing to the introduction of many false equations to which the " secular equation " has been applied.

Ptolemy's eclipses range from B.C. 720 to A.D. 136 and are all given with commendable accuracy. Those of Confucius, which are contained in his historical work the Chun-Tsiu, extend from B.C. 719 to B.C. 494.

The eclipse of Thales which has a singular historical setting as it " put an end to the war between the Medes and Lydians," according to Herodotus, for it happened when the battle was at its heat, and the day was suddenly turned into night. There was a total eclipse of the Sun over Asia Minor on May 28th, B.C. 584, and this has been accepted as the day by Pliny, Newton, Ferguson and others. It has been suggested that Thales was able to predict this eclipse because there had been one 18 years previously in B.C. 602 (astron.). There is nothing against this suggestion, but it shows that Thales knew something more than was contained in the knowledge of the Saros, since he was able to speak of it as visible in those parts which, of course, does not follow from the mere fact that eclipses recur after 223 lunations.

Other eclipses of interest are those of Larissa, mentioned by Xenophon in his Anabasis III-4, and that of Xerxes. The former took place at Nineveh on May 19th, B.C. 556, and was total, and the latter took place on Feb. 17th, B.C. 477, and was nearly total. The eclipse of Thucydides which occurred during the Peloponnesian War, which Anaxagoras is said to have predicted, was seen at Athens on August 3rd, B.C. 430. It deserves mention inasmuch as Sir J. Herschel, speaking of it, says that it was total, whereas it was an annular eclipse. This remark of the great astronomer might easily arise from the observation that the Sun was within 5 degrees of the Node, whereas a total eclipse can only happen under these conditions when the semi-diameter of the Moon is greater than that of the Sun, and this latter may not have entered into the astronomer's calculation at the moment.

A far more remarkable error is that of the French Connaissance de Temps, 1912, which gave the eclipse of April 17th in that year as total, whereas it was only annular, as stated by the Nautical Almanac. Here, again, there may have been a misapprehension of the relative apparent diameters of the Sun and Moon at the time.

The eclipse of April, 1912, is remarkable from an historical point of view, because it immediately preceded the outbreak of the Balkan War, which was destined to be the precursor of the greatest war that humanity has ever engaged in, and which, directly connected with the Balkan War, had its origin in the Slav and Austrian hostility.

CHAPTER V.

To Calculate an Eclipse of the Sun.

FOR this purpose it will be necessary to have a set of Tables which are based on the observed average or mean length of a lunation, and which in addition thereto give the progress of the Sun's longitude, the Sun's anomaly, the Moon's anomaly, and the distance of the Sun from the Node.

A very useful set of Tables will be found in Ferguson's "Astronomy," which are based on Dr. Pound's estimates of the mean values for a lunation. These give the dates of the New Moons in March for successive years, the Sun's mean anomaly, the Moon's mean anomaly, and the Sun's distance from the Node, and therefore contain all that is required for the estimate of the mean time of a lunation for any month or year. But also there are five principal equations from which we can get the true time of the said lunation. These are: (1) The equation known as the annual equation, depending on the Sun's anomaly; (2) the equation of the Moon's anomaly due to the Sun's anomaly; (3) the equation of the time of syzygy due to the Moon's anomaly; (4) the equation of time due to the difference between the anomalies of the Sun and Moon, and therefore of their relative places in their respective orbits; and (5) the equation due to the Sun's distance from the Node.

All other Tables are constructed on the same

lines, and only differ in the values of the several equations.

The Tables I am using are those which I consider to be the best. They are based on the observed mean value of a lunation = 29d. 12h. 44m. 2.8464s., which is rather shorter by about one-fifth of a second of time, and to these I apply the secular equation for centuries past.

Having the date of any one lunation in March, the mean values for successive lunations in the same month over a period of 100 years are given, and also the multiplied values for periods of one, two, three hundred up to a thousand or more years. From these it is possible to obtain the values for any year, past or future.

The rules of procedure are as follows :--

1. Consider whether the given date is a secular or astronomical one. If secular, subtract one year for all years before the era A.D. This will reduce it to the astronomical equivalent. Thus 763 B.C. = 762 astronomical.

2. Consider whether the date be Old Style or New Style. If Old Style, that is, the style current before the year A.D. 1753, this must be retained when the Tables are set for Old Style. Otherwise a correction is necessary to convert it to New Style. This is effected in the following manner :—

(a) Divide the complete centuries of the date by 4 and throw out the remainder.

(b) Multiply the dividend by 3 and cast in the above remainder.

(c) From the result take 2, and the remainder will be the number of days to be added to Old Style to get the New Style, or subtracted from New Style to get the Old Style.

3. If the given year is not included in the Tables

of Mean Lunations, then add such centuries of years to the given date as shall bring it to one that is included in the Tables. Note the number of centuries required for this purpose.

4. For the equivalent year found in the Tables, take out the mean New Moon in March, the Sun's distance from the Node, the Moon's mean anomaly, and the Sun's mean anomaly.

5. Subtract therefrom (for past centuries) or add thereto (for future centuries) the values for the number of centuries between the date in the Tables and the date of the eclipse as determined under Rule 4. If a subtraction cannot be made, as when the century values bring out a date in March which is later than the date in March given in the Tables, then add the values for one lunation, and also the values due for the same to the Sun's distance from the Node, the Moon's and Sun's anomalies, and make subtraction thereafter for the values for complete centuries.

6. Having thus the data for the mean lunation in March in the required year (B.C. or A.D.), you will now correct these values by the addition of so many lunations as shall bring it to the month in which the eclipse took place or will take place: as if in April, the values for 1 lunation must be added, for May 2 lunations, for June 3 lunations, and so on.

7. The next step is to correct these mean values by the following equations:—

(a) The annual equation. This is a correction of time due to the Sun's anomaly. The amount is nil at the apogee and perigee, and amounts to 4h. 10m. at three signs from the apogee.

(b) The correction of the Moon's anomaly. This is due to the Sun's anomaly, and amounts to $1^\circ\ 35'$

at the maximum when the Sun's anomaly is three or nine signs.

(c) The lunar equation. This is a correction of time due to the Moon's corrected anomaly, and amounts to 9h. 48m. at the maximum, the Moon being then three or nine signs from the apogee.

(d) The correction of the Sun's distance from the Node. This is due to the Sun's mean anomaly, and may amount to 2° 4′.

(e) The anomalistic equation. This is due to the difference between the Moon's anomaly and the Sun's anomaly, or A – a. This may amount to 2m. 47s. of time.

(f) The reduction. This is due to the Sun's distance from the Node as corrected under Rule (d). This may amount to 15m. of time.

There remains the true time of the ecliptic conjunction for the required date.

It will be seen that there are only two main equations of the mean time of a lunation. The first is that due to the Sun, the second that due to the Moon. But before we can extract the latter we have to correct the Moon's anomaly. There are also two minor equations of the mean time. The first is due to the difference of the anomalies of the Sun and Moon, and the other to the Sun's distance from the Node. But before we can apply the latter we have to correct it from the mean to the true distance. The whole process is extremely simple, and amounts to no more than mere addition and subtraction when once the Tables for these equations are in hand.

We may now take two examples of eclipses of the Sun, one for a date in the past, another in the future. For this purpose we shall employ the Tables given in my most recent work.

Solar Eclipse, June, B.C. 762.

The year is astronomical, the secular equivalent being B.C. 763. The month and day of the month will be in Old Style, as our Tables are set for that style.

Year.	March.	Sun-Node.	Moon's Anom. = A.	Sun's Anom. = *a*.
	D. H. M.	S. ° '	S. ° '	S. ° '
1738 ...	9 1 15	6 29 40	10 6 24	8 19 30
1*l*	29 12 44	1 0 40	0 25 49	0 29 6
	38 13 59	8 0 20	11 2 13	9 18 36
—2500 ...	19 20 35	5 4 34	6 9 59	11 25 39
B.C. 762	18 17 24	2 25 46	4 22 14	9 22 57
+ 3*l* ...	88 14 12	3 3 1	2 17 27	2 27 19
	107 7 36	5 27 47	7 9 41	0 20 16
a	— 1 24	— 42	— 32	
	107 6 12	5 27 5	7 9 9A'	
A'	— 5 49	(S—N)	0 20 15*a*	
	107 0 23		6 18 53	
A'—*a* ...	— 1		(A'—*a*)	
	107 0 22			
S—N ...	— 1			
	107 0 21	=June 15th	, 12.21 p.m.	
Sec. Eqtr.	— 3 13			
	9 8	a.m.		
E. long....	2 52	Nineveh		
	12 0=	Noon at Ni	neveh	

The Sun's longitude was ♊ 15° 44' and its declination 22° 22' N. The Moon was 2° 55' from the Node, and its latitude about 15' N. Its horizontal parallax was about 60', and this gives the angle of apparition about 14° 28', which, added to the N. declination 22° 22', gives the latitude of the place where the eclipse was central = 37° N.

By Rule 3 I find the mean values from the Tables for March, A.D. 1738, and from these I take the values for 2,500 years, because 762 added to 1,738 makes a complete number of centuries, namely, 25.

A glance at these values shows that we must take out the values for 1738 (Rule 4) and add one lunation, because the days in March are only 9 against 1,738 while those against 2,500 years are 19 (Rule 5). By this process we derive the values given in the Table on the preceding page.

This is close on the latitude of Nineveh, and as the lunation took place at noon in that locality, the eclipse must have been immediately overhead, and as the Moon's diameter was greater than that of the Sun, the eclipse was total. No wonder, then, it became the subject of a special record on the Assyrian Tablets, which fortunately have been preserved to us.

We may now take an eclipse which, from the epoch of our Tables, is to be regarded as a future eclipse. It will be found in the Nautical Almanac for the year 1916, on February 3rd.

ECLIPSE OF THE SUN, FEBRUARY 3RD, 1916.

The date here is New Style, and we therefore deduct 13 days, as our Tables are for Old Style. This gives us January 21st (O.S.). The following is the calculation :—

Year.	March.	S—N.	A	*a*.
	D. H. M.	S. ° ′	S. ° ′	S. ° ′
1715 ...	23 8 45	4 19 14	5 24 26	9 3 35
+ 10*l* ...	295 7 20	10 6 42	8 18 10	9 21 3
1716, Jan.	318 16 5	2 25 56	2 12 36	6 24 38
+ 200 yrs.	8 16 13	9 8 56	5 1 0	0 6 38
1916 ...	327 8 18	0 4 52	7 13 36	7 1 16
a	+ 2 12	+ 1 5	+ 50	
	327 10 30	0 5 57	7 14 26A	
A	— 6 29		7 1 16*a*	
	327 4 1		0 13 10=	
A—*a* ...	+ 1		(A—*a*)	
S—N ...	+ 3			
	327 4 5=	Jan. 21 (O.S)	4.5 p.m. =	Feb. 3 (N.S.)

The secular equation is about one minute, which would bring the time up to 4.6 p.m. at Greenwich.

At its distance from the Node the Moon would have about 30′ N. latitude, and its anomaly shows the horizontal parallax to be 60′ 10″, which gives the angle of apparition =30°, and the Sun's declination being about 17° S., the eclipse would be central in latitude 13° North. The time of the lunation gives an hour angle of 4h. 6m., equal to 61° of West longitude, so that the Sun will be eclipsed when overhead at St. Vincent Isle in the Antilles. The Moon's diameter being greater than that of the Sun as 16′ 24″ is to 16′ 15″, the eclipse will be total.

These two examples of solar eclipses will enable the student to follow out the use of the Tables and compute other eclipses at pleasure. The chronological value of eclipses cannot be over-estimated, and it may in truth be said that chronology never had a sound basis until it was linked to astronomy. Classical history has been rescued from the region of romance by the fact that eclipses mentioned by various authors in connection with certain events have since been verified, both in regard to the time and place of their happening. In the course of these pages I propose to invest eclipses with an interest of quite another nature.

CHAPTER VI.

To Calculate a Lunar Eclipse.

THE procedure is in every respect the same as that taken for calculating an eclipse of the Sun, but the addition of half a lunation is necessary to bring it to the Full Moon, the Tables being set for New Moons in March.

But whereas an eclipse of the Sun will last only about two hours, an eclipse of the Moon may last for several hours. This is due to the fact that in the first instance we are concerned only with the relative apparent diameters of the Sun and Moon and their relative motions, while in the second case we are dealing with the Moon's progress through a sector of the Earth's shadow, which is always greater than the apparent diameters of the Sun and Moon combined.

Suppose the Sun to be always throwing the shadow of the Earth into space. This shadow will take the form of a cone, which at the Earth's surface will be of the same diameter as the Earth, but at a distance from it it will be proportionately less, until at last it disappears in space. Now as the Moon is at a distance from the Earth, it will, in its course about the Earth, cut into the shadow only at such times as it comes to the Full when within a certain distance of the Node—*i.e.*, the point in its orbit at which the Moon, Earth, and Sun are all in direct line with one another. And its distance from the Earth at that time will be determined by the place in its orbit, and its consequent distance from the Earth.

For we have already seen that the apparent motion of the Moon about the Earth is elliptical, so that at one time it is in apogee (furthest from the Earth) and at another time in perigee (nearest the Earth). And just as the apparent size of the Moon is controlled by its distance from us at any moment of time, so the Earth's size from the point of view of the Moon is controlled by the same factor. Thus when at its perigee the Moon would see the Earth to be larger than when at its apogee. This size of the Earth is measured by the Moon's horizontal parallax, which is half the apparent dimension of the Earth as seen from the Moon.

All apparent dimensions are measured by the minutes and seconds of space they appear to occupy. Thus the mean apparent diameter of the Sun is 32′ 2″, and of the Moon 31′ 16″, and for purposes of calculation we employ half these amounts, namely, 16′ 1″ and 15′ 38″, which are the semidiameters of the Sun and Moon. The Earth's mean diameter as seen from the Moon is 114′ 34″, and its semidiameter 57′ 17″. Therefore if we curtail the apparent semidiameter of the Earth by the Sun's semidiameter and add that of the Earth as seen from the Sun, we shall have the half dimension of the shadow through which the Moon passes at its eclipse. The rule is : To the parallax of the Moon add the parallax of the Sun (always taken as 9″) and subtract the semidiameter of the Sun. There remains the diameter of the cone or dark shadow of the Earth at that point through which the Moon's orbit passes. Then the rate at which the Moon overtakes the Sun will determine the time it takes in traversing this section of the cone.

In order to illustrate these remarks we may have recourse to one or two illustrations, one of which may be an historical eclipse concerning which there is no doubt whatsoever. This is the large Babylonian eclipse which was recorded by Hipparchus, and from him received and transmitted to us by Claudius Ptolemy. It is of interest here to note that Hipparchus made certain observations of the longitude of stars in his day, and by comparing them with the observations of his predecessors he discovered that the distances of the stars from the ecliptic and from each other were invariable, but that their longitudes and declinations were changed in the course of time. By comparing his own observations with those of Timocharis, he calculated that they increased their longitude by 1° in 100 years. He had discovered the precession of the equinoxes, a fact which was destined to re-model the whole of our astronomical ideas!

Total Lunar Eclipse, B.C. 721, March 19th.

Here the date B.C. 721 is secular chronology, and the astronomical year is one less, namely, B.C. 720. March 19th is Old Style, and our Tables being adjusted to that style, there will not be any change necessary. By the addition of 25 centuries we obtain the year A.D. 1780, since B.C. 720 + A.D. 1780 = 2500. We have therefore to find the Full Moon in March, A.D. 1780, and take from it the values for 2,500 years, and we shall have those for B.C. 720, March Full Moon. This will be subject to a secular equation, on account of the known acceleration of the Moon in its orbit being greater then than now, our Tables being framed on the mean value of a modern lunation.

Year.	March.	S—N.	A	*a*
	D. H. M.	S. ° ′	S. ° ′	S. ° ′
1780 ...	23 23 2	10 18 21	1 21 8	9 4 18
½*l*	14 18 22	0 15 20	6 12 54	0 14 33
Full Moon	38 17 24	11 3 41	8 4 2	9 18 51
—2500yrs.	19 20 35	5 4 34	6 9 59	11 25 39
March ...	18 20 49	5 29 7	1 24 3	9 23 12
a	+ 3 34	+ 1 54	+ 1 27	
	19 0 23	6 1 1	1 25 30A′	9 23 12
A′... ...	+ 8 24	(S—N)	9 23 12*a*	
	19 8 47		4 2 18	
A′—*a* ...	+ 2		(A′—*a*)	
S.—N. ...	+ 0			
	19 8 49			
Sec. Eqtr.	— 3 7			
	19 5 42			
E. long....	+ 2 58			
Babylon...	19 8 40	=March 19,	8.40 p.m.,	B.C. 720

The anomaly of the Moon being 1s. 25° 30′, its horizontal parallax was 55′ 15″. The Sun's hor. par. is always taken at 9″, and this added to the Moon's will make 55′ 24″. The Sun's semidiameter as derived from its anomaly was 15′ 54″, which, taken from 55′ 24″, leaves 39′ 30″ as the semidiameter of the dark cone at the Moon's orbit.

To find the time of the semiduration of the eclipse we compare the Moon's horary motion 31′ 6″ with the Sun's horary motion 2′ 26″, according to their anomalies, and the difference is 28′ 40″, at which rate the Moon overtakes the Sun, and so traverses the Earth's shadow, which moves always against the Sun. And as we are dealing only with the semidiameter of the Earth's shadow, the time will be half the duration of the eclipse. Thus: As 28′ 40″ is to 60m., so is 39′ 30″ to about 82m. The half duration will therefore be 1h. 22m., and the eclipse, which was central at 8.40 p.m., would begin at about 6.18 p.m. at Babylon and last till 10 o'clock. The Sun being 6s. 1° 1′ from the Ascending Node, the Moon must have been in direct opposition, and therefore in 0s. 1° 1′ from the Ascending Node and in N. latitude. Its visible path with the ecliptic being (always) 5° 35′ at eclipses, the Moon would enter the shadow from the right below and leave it on the left above, as here shown in three phases in Fig. 5.

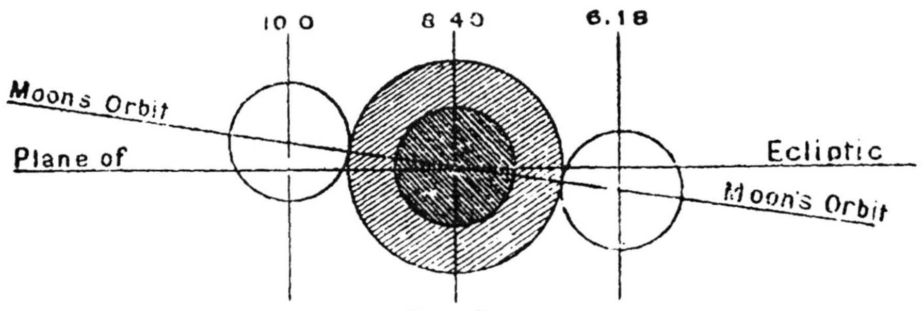

Fig 5.

In this sketch the Moon is made to form the opposition when exactly on the Node. This was not the case in the above eclipse, the Moon having passed the Node by a whole degree, and it therefore had some N. latitude when at the middle of eclipse. The Moon's latitude would be a little over 5′, as shown by its distance from the Node = 1° 1′. The sun being in Pisces, the Moon would be in the sign Virgo, and this was the first of two eclipses in that sign which were held to be portents of the fall of the Babylonian Empire.

Partial Eclipse of the Moon, July 15, 1916.

The date here is New Style, and we have therefore to subtract 13 days to bring it to Old Style, in conformity with our Tables.

Year.	March.	S—N.	A	*a*
	D. H. M.	S. ° ′	S. ° ′	S. ° ′
1716 ...	11 17 33	4 27 17	4 4 14	8 22 51
200 ...	8 16 13	9 8 56	5 1 0	0 6 38
1916 ...	20 9 46	2 6 13	9 5 14	8 29 29
4*l*	118 2 56	4 2 41	3 13 16	3 26 25
July =	138 12 42	6 8 54	0 18 30	0 25 54
— ½*l* ...	14 18 22	0 15 20	6 12 54	0 14 33
	123 18 20	5 23 34	6 5 36	0 11 21
a	— 0 48	— 23	— 18	
	17 32	5 23 11	6 5 18A′	
A′... ...	— 51	(S—N)	0 11 21*a*	
	16 41		5 23 57	
(A—*a*) =	0		(A′—*a*)	
S—N ...	— 4			
	16 37			
Sec. Eqtr.	+ 2			
	16 39	= July 1,	(O.S.) 16h.	39m.
		July 14,	(N.S.) 16h.	39m.
		= July 15,	4.39 a.m.	

The eclipse is thus seen to take place from the ecliptic opposition of the luminaries at 4.39 p.m. on July 15th, 1916. The Moon's anomaly gives its parallax at 61′ 20″ and its apparent semidiameter 16′ 45″, and its horary motion 38′ 9″. The Sun's anomaly shows its horary motion to be 2′ 33″, and its parallax will, as usual, be 9″. The sum of the parallaxes is therefore 61′ 29″, from which we take the Sun's semidiameter 15′ 46″ and obtain the apparent semidiameter of the Earth's shadow at that part of the Moon's orbit where it crosses the dark cone—namely, 45′ 43″. The difference of the horary motions of the Sun and Moon is 38′ 9″ − 2′ 23″ = 35′ 46″, at which rate the Moon will overtake the Sun per hour. Then as 35′ 46″ : 60m. : : 45′ 43″ to 76m. nearly. Hence the Moon will enter the dark cone of the Earth's shadow 1h. 16m. before the time of ecliptic conjunction—namely, at 3.23 a.m.—the middle of the eclipse will be about 4.39, and the end about 5.55 a.m. As the Sun will be under the Earth only till 4 a.m. the eclipse of the Moon will not be visible on this side of the Earth. The place of the maximum phase will be equal to the hour-angle of the Moon. The Sun's distance from the meridian of Greenwich would be, at 4.39 a.m., about 110° E., and if we take this from 180 we shall have the place of the Moon 70° West longitude, the first contact with the shadow will take place when the Moon is on the meridian in about 50° West longitude, and the last contact when the Moon is on the meridian in about 90° W. The eclipse would nowhere be total, for the Moon is seen to be 6° 49′ from the Node at the time of opposition, and hence we have the dimensions of the eclipse of about 9 digits only.

The Sun being 6° 49′ from the descending Node, the Moon at the opposite point would be the same distance from the ascending Node, and therefore in S. latitude 0° 35′ decreasing. The following is the figure for the eclipse (Fig. 6). The latitude of the Moon being S., the disc will be eclipsed on the North.

We are now in a position to turn to an entirely new aspect of the subject. We have seen under what conditions an eclipse of the Sun or Moon is bound to take place, and we have shown the means by which the dimensions, duration, and locality of

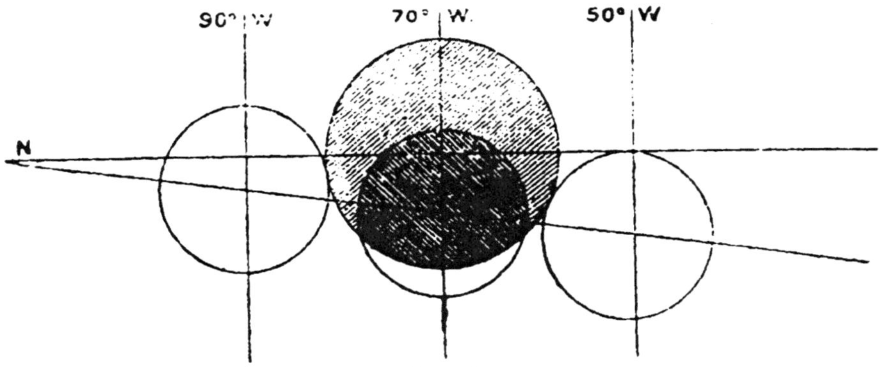

Fig 6

the eclipse may be estimated. These are all natural astronomical facts brought under mathematical consideration. But all natural phenomena are symbolical. The whole universe is the expression of an Intelligence, which, by reason of its mathematical integrity, is rendered to us intelligible. We have to consider this symbolism in some of its more conspicuous details and endeavour to find a system of interpretation which, when brought into line with other branches of natural physiognomy, may be held to constitute a system of Cosmic Symbolism. In other words, we have to give a human value to our astronomy.

CHAPTER VII.

Eclipse Signs.

DOUBTLESS it would not occur to the scientific mind, concerned as it is only with the phenomenal appearances of nature, to suspect that the part of the zodiac in which a luminary is eclipsed has any degree of significance. To the mind gifted with a certain degree of " scientific imagination," that faculty so highly extolled by the late Professor Huxley, the position would present itself quite differently, however. To a man afflicted with blindness, it does not signify much that the Sun is eclipsed at high noon or that the electric lighting gives out just at the moment that some spectacular exhibit is about to begin. It will depend entirely on his education and training as to whether he is able to distinguish mentally between these two kinds of light. But although a scientific man would be able to recognise the difference between a transit of Venus over the Sun's disc and an annular eclipse of the Sun, both essentially belonging to the same order of phenomena, it is extremely problematical whether he would be in a position to recognise the difference of their symbolical values, or even predicate the physical effects of the two happenings. Except in some remarkable instances of men with a genius for scientific exploration, men of science have not disposed themselves to trace any terrestrial effects due to celestial phenomena, with the sole exception, perhaps, of the action of the luminaries on the tides.

But I think we may be able to show, not only that there are very definite effects due to the phenomena of eclipses as distinguished from lunations, but also that these effects are modified by the part of the heavens from which they may be said to operate. Then, the integrity of Nature being admitted, and man's dependence on his greater environment being established, it should be a matter quite within the domain of a scientific system to trace the connection between the terrestrial phenomena as effects, and the significance of these phenomena as portents and symbols in the human mind. The validity of the symbol will depend largely upon the implied connection between the celestial and terrestrial phenomena, and the consequent meaning that we import to them. But in this place it will be sufficient if we show that a definite sequence of events are directly attributable to eclipse influence, and for obvious reasons these will be better appreciated as physical phenomena than as mental and psychological, although the connection between the two orders will be apparent to most minds.

For purposes of distinction the signs of the zodiac are divided into four triplicities and three quadratures. The triplicities are named after the four states of matter : Fire, Air, Water, and Earth. The signs in each triplicity are 120°, or four signs apart. Thus Aries, Leo, and Sagittarius are fiery signs; Taurus, Virgo, and Capricorn are earthy signs; Gemini, Libra, and Aquarius are airy signs; Cancer, Scorpio, and Pisces are watery signs.

These are referred to as the " elemental " natures of the signs.

The quadratures are Cardinal, Mutable, and

Fixed. They are classified thus: Aries, Cancer, Libra, and Capricorn are cardinal; Taurus, Leo, Scorpio, and Aquarius are fixed; Gemini, Virgo, Sagittarius, and Pisces are mutable signs. These are referred to as the "constitutional" natures of the various signs.

There are thus three constitutions of the four elements.

From most ancient times there have been four particular signs regarded as "corner-stones" or "pillars of the corner" in the great Cosmic Temple. They have entered into the construction of the sacred monuments of all nations. These are the four fixed signs represented by the Bull, Lion, Serpent, and Man, denoted by the signs Taurus, Leo, Scorpio, and Aquarius. The first, second, and fourth of these have remained without alteration, but the third of them, Scorpio, has successively been associated with the Dragon, the Serpent, and the Eagle. A brief statement of some of the myths surrounding these signs will help to make the transition clear.

In ancient China they recognised four quarters of the year, and these they characterised by the culminating constellations at sunset. They were the "Red Bird," answering to the sign Leo; the "Blue Dragon," or the sign Scorpio; the "Black Warrior," or sign Aquarius; and the "White Tiger," or the sign Taurus. Here we have a direct identification of the Man with Aquarius and of the Dragon with Scorpio. With the Hindus the sign Taurus is the chariot of Vishnu, symbolised by the Bull. The tradition has come down to us in the legend of St. George and the Dragon, wherein the sign Taurus is St. George and Scorpio is the Dragon. Taurus is the Bull, the ancient beast of agriculture, the earth-worker, and George means "a farmer," from Geo-urgon (earth-worker). We find the Bull in use on the fields and on the threshing floor and in the shafts. In the East it is still largely employed in agricultural work owing to its great strength and power of endurance, and in parts where they do not kill for food it is considered more serviceable than the horse for agricultural work. The entry of the Sun into the sign Taurus was celebrated among the Hindus and Egyptians as the feast of Maya or Maia, from which we have derived our May Day festival. At such times the ancients decorated the White Bull and placed between its horns a golden disc, symbol of the Sun in the sign Taurus, and this was followed by a procession of virgins in token of the fecundity of Nature springing into new life at this season of the year. We find the four symbols of the fixed signs entering into the construction of the Assyrian Bull, which had the head of a Man, the wings of an Eagle, and fore and hind parts of the Bull and Lion respectively. Also we find it again in the Sphinx and in the figure of the Cherubim in the Hebrew religious system. Everywhere these four signs have been regarded with exceptional distinction, and the reason is that these four signs at one time corresponded astronomically with the constellations of the same names and marked the equinoxes and solstices. And this fact gives us the period of their evolution as symbols in human thought. Allowing 2,160 years as the mean value of the precession of the equinoxes through a sign, they being now at the end of Aquarius and Leo, we find the passage through Pisces and Aries to extend over 4,320 years, and half of the sign Taurus would bring us to the middle of that constellation about the 5,400th year from the present date, namely, about 3500 B.C., and possibly we may go back to the year 4580 B.C. as the beginning of this conception regarding the foundation signs. Scorpionis, Serpentarius, Draconis are all linked together in this concept with the constellation Aquilla, the Eagle. The twelve tribes of Israel answer to the twelve signs or constellations, and in this scheme we have—

Reuben.—Taurus, " the first-born."
Simeon.—Gemini } " the twins."
Levi.—Cancer }
Judah.—Leo, "the lion."
Zebulon.—Virgo, " the harbour."
Issachar.—Libra, " the servant to tribute."
Dan.—Scorpio, " the serpent in the path."
Gad.—Sagittarius, " the horseman."
Asher.—Capricorn, " the pleasure of kings."
Naphtali.—Aquarius, " the spreading oak."
Joseph.—Pisces, " the fruitful vine."
Benjamin.—Aries, " the ravening wolf."

Here Taurus is the first-born, because it is the first sign of the ancient zodiac as received by the Patriarchs and Shepherd Kings. Leo is the lion, Scorpio is the serpent, and in its association with the tribe of Dan it should be observed that the symbol was replaced on the banner of that tribe by an eagle, which arose from the idea that Scorpio was " the accursed sign," and Aquilla is the constellation that rises with Scorpionis. Aquarius is here the " spreading oak," and in the old planisphere there is seen the "tree" figured in connection with the sign Aquarius. Instead of the common translation, " Naphtali is a hind let loose, he giveth goodly words," we have the more accurate rendering: " Naphtali is a spreading oak who yieldeth goodly branches," the word *imri* meaning not only words, but branches, rays, or emanations. Thus in the 19th Psalm of David it is said of the celestial bodies (*hashemayim*), "Their line (*qenon* = rule, dominion) hath gone out (extended) to all the earth, and their rays (*imri*) to the uttermost parts thereof."

Now concerning these four fixed signs of the zodiac, it is said that when there shall be a majority of the planets, including the luminaries, in these signs, or when there shall be eclipses taking place in them followed by the transit of planets over the places of such eclipse, violent earthquakes are to be expected in those parts of the world that are liable to them. And it has more recently been observed that to natural seismic disturbances we are justified in adding explosions in coal mines and pits, also submarine earthquakes producing tidal waves. For then it would appear there is an assault upon the foundations of the Earth as represented by the foundation signs Taurus, Scorpio, Leo, and Aquarius, which, it will be observed, are all in quadrature or opposition to one another.

Let us see if this is borne out by our experience. The last great earthquake took place in Abruzzi, Italy, in the month of February, 1915, when it will be seen that the Node, the Sun, Uranus, Jupiter, and Mercury were all in the sign of Aquarius.

On April 22nd, 1902, there was an eclipse of the Moon in the sign Scorpio, which was total, the Moon being within 3° of the ascending Node in the sign Scorpio. On May 7th at the next lunation there was a solar eclipse in Taurus, the planets Mars and Mercury being also in the same sign as the luminaries. The Sun had 16° N. declination, and on April 22nd its declination was 12° N. The luminaries therefore passed directly over the islands of the Antilles, and it was at Martinique on the very day of this conjunction in Taurus that Mont Pelée burst into eruption and devastated the whole island, with immense loss of life and property. We have here an object lesson of the greatest interest. Not only do we find the Sun and Moon both pulling together for some hours before and after the actual time of eclipse, but also the Moon close to its perigee or nearest distance from the Earth, and affecting a sporad of islands which are known to be highly volcanic. The action of the Moon upon the tides is well known and understood, and it is obvious that its action must be greatest when the Moon is in perigee at the time of its conjunction with the Sun. We have every reason to presume that the same force that lifts a vast mass of water to a height of several feet in the course of a few hours is capable of exerting a proportionate action upon the fluidic contents of the earth, and that the action of the luminaries does not cease at the earth's surface. Indeed, it may be said that the phenomenon of the neap tide, when the Sun and Moon are pulling in opposite directions upon the Earth's mass, can only be explained on the supposition that the Moon not only draws the ocean waters, but also the body of the Earth itself. For if the Sun has any degree of tidal influence equal or superior to that of the Moon we should have the phenomenon of a tide every day at noon. But such is not the case. And in regard to the action of an eclipse upon the earth, it is to be observed that the rapid obscuration of the Sun, and the cutting off of its rays locally, cannot but result in a very rapid fall of temperature. This would be compensated by an uprush of heat from the Earth's interior to restore equilibrium, and if that part of the Earth's crust is already weakened by volcanic action or other cause of attrition, the strain could not be sustained, and an earthquake is exactly what we should expect.

But it is probable that we need not look always to find the eclipsed luminary in a fixed sign of the zodiac. What appears more reasonable is that we should look to the coincidence of an eclipse with some area of the Earth's surface which is included in what are recognised as "earthquake or volcanic areas." These are easily recognised by either the existence of an active or slumbering volcano, or the existence of a group of detached islands or a sporad such as mark the archipelagos. These may be found in the line of islands running from Kamchatka southward to the Corea and enclosing the Sea of Okhotsk and the Sea of Japan; in the islands of the Pacific terminating in New Zealand; in the Malay Archipelago and the Philippines; in the Grecian Archipelago; in the West Indies, enclosing the Gulf of Mexico and the Caribbean Sea; and in the northerly islands known as Franz Josef Land, Spitzbergen, and the Parry Islands.

These may be regarded as constitutional weak points in the Earth, and subject to seismic disturbance upon the least strain. But there may be others, and any area in which there is temporary strain may be thus affected. Thus in the year 1693 the eclipses were falling in the signs Sagittarius and Gemini, and happened to coincide with the opposition of the major planets Uranus and Saturn, which were then in Gemini and Sagittarius, and therefore acting in the same line as the Sun and Moon at their ecliptic conjunction and opposition in those signs. Among the more remarkable results of this concatenation of celestial influences we may note the great earthquake which took place in Sicily, at Catania, when no less than 100,000 people perished.

But we must not regard eclipses as solely connected with earthquakes. They may be reasonably connected with all sorts of other mundane events if we come to regard them as symbolical portents and not merely as physical causes. And this seems the more satisfactory line of procedure, for we continually find that predictions have been made in regard to eclipses which could not be argued out upon merely physical premisses. There remains the fact that earthquakes have been predicted with considerable accuracy, both as to time and place of their occurrence. I may here cite an instance of a prediction by Commander Morrison, R.N., who wrote concerning the eclipse of the Sun on 6th June, 1853, as follows: As Mars and Saturn are in the sign Taurus in the precedent angle of the eclipse at Panama, I have no doubt there will be a fearful amount of earthquakes there, and all about the isthmus of Darien, the shocks extending to Carthagena, along the northern coast of South America to Honduras, California, Florida, and the West Indies. These events may be looked for, among other periods, in July, 1853, about the 16th day.

On July 15th, at 2.15 p.m., at Cumana, which is on the northern coast of South America, a terrible earthquake occurred, in which thousands of lives were lost, and not a family escaped without some loss of its members. This prediction was written and published a full year before its fulfilment. It deserves to be kept on record as a luminous case of scientific prediction. The reader will observe that considerable importance attaches to the conjunction of the planets Saturn and Mars in Taurus coinciding with the eclipse, but the seismic effect is most clearly pinned down to the locality in which the eclipse was visible, and almost overhead at the time of centrality.

The same writer declared that the sign Gemini was the ascendant of the City of London, and that the 18th degree more particularly was connected with the affairs of the Metropolis. In such case the 13th degree of Aquarius would be on the mid-heaven, and in this connection the following excerpt from the almanac by Saunders, published in the year 1664, will be of interest. Referring to the recent conjunction of Jupiter and Saturn in opposition to the ascendant of London and the eclipse which fell on the meridian of the same horoscope in Aquarius 14, the writer says :—

"Eclipse of the Moon, July, 1664. This eclipse falls in the 14th degree of the sign Aquarius, and is very formidable, and threatens much sorrow to the western world. This is the very radical position of the Moon in the Radix of London and in the tenth angle. It hath been the subject of my inquisitions, and I find that for hundreds of years past, that never any plague, contagious disease, fire, murther, sedition, conspiracy, or tumultuous mischiefs, but hapned either when some eminent conjunction, notable eclipses, or other malevolent position or transits were made in or near the angles of the City Radix, and therefore the preceding conjunction of Saturn and Jupiter and in Sagittarius 13, and this eclipse in Aquarius 13 do *prodigiously threaten London with raging sickness* and other mischiefs which I prayed God to avert." This remarkable prediction was immediately followed by the outbreak of the plague in London. But the author further cites the following eclipse of the Moon in

January 1665, which fell in Leo 12, on the degree held by the planet Mars in the City Radix, and in opposition to the mid-heaven of that horoscope, which apparently signified the Fire of London in the following year. These predictions seem to give a purely symbolical value to eclipse positions in distinction from any part they may play as physical causes in Nature, for it cannot by any means be argued that these configurations "caused" the Great Plague in the one case or the Great Fire in the other. The belief that the celestial bodies may have an othic value as portents and signals is not contrary either to reason or religious principle. In the belief of the ancient sages the celestial bodies were intended to serve as indicators, pointers, and signals: "Let them be for signs (*othuth*) and for seasons, for days, and for years," being the clear mandate. The most abundant proof exists that this is an intelligent conception of the intention and purpose of the Divine Mind, for modern experience accords entirely with the views of the Chaldeans and Hebrews. As an instance we may cite the most recent conjunction of Saturn and Mars in the middle of the sign Gemini, which occurred in August, 1913, and this was the first time these planets were conjoined in that part of the zodiac for 265 years. The invasion of London and the institution of the Commonwealth in 1649 is exactly 265 years from the outbreak of the Great War in 1914, when London again became the objective of an invading force. And in this connection I find two eclipses of the Moon, one in 1648, on June 5th (O.S.), in Sagittarius 23, and the other on November 29th (O.S.), with the Moon in Gemini 17, close to the ascendant of London. At the outbreak of the War in 1914 we find the planet Uranus on the mid-heaven of London. What are called the "human" signs in astrology are Gemini, Sagittarius, Virgo, and Aquarius, to which some writers add Libra also. The bestial signs are Aries, Taurus, Leo, the first half of Sagittarius, and Capricornus. The aquatic signs are Cancer, Scorpio, and Pisces. These observations will give a preliminary view of the method of interpretation followed by students of horoscopy and astrological science. It should here be remarked that the "total" eclipse of the Moon referred to by Saunders took place on July 27th (O.S.), at 11.2 p.m., and was therefore close to the meridian of London, while its longitude was Aquarius 13. The reader will be able to calculate the elements for himself from the Tables. The eclipse is not in the Catalogue of Ricciolus, and so far as it known, Saunders is the only publisher of this important event.*

---

* Since writing the above I have been informed by the Astronomer Royal that "the eclipse was predicted with full details and a diagram in Vincent Wing's Ephemeris for 1664."

CHAPTER VIII.

Eclipse Indications.

THE old writers who have delivered to us the doctrine of eclipses in a more or less fragmentary and detached form, have uniformly made distinction of the particular decanate in which the eclipsed luminary is placed.

The signs are divided into three equal parts of 10° each, and these are called "decans" or decanates. A considerable difference arises between the Oriental and Occidental systems in regard to the signature or symbol attaching to these decans. In the East we find that the rulership of the successive decans follows the order of the Triplicities, while in the West the order follows the Chaldean enumeration, beginning with the first decan of Aries under Mars' rulership. The two systems ought therefore to be compared and subjected to criticism. They stand as follows :—

ORIENTAL SYSTEM.

Aries	...Aries,	Leo,	Sagittarius	...Mars, Sun, Jupiter.
	10°	20°	30°	
Taurus	...Taurus,	Virgo,	Capricorn	...Venus, Mercury, Saturn.
	10°	20°	30°	
Gemini	...Gemini,	Libra,	Aquarius	...Mercury, Venus, Saturn.
	10°	20°	30°	
Cancer	...Cancer,	Scorpio,	Pisces	...Moon, Mars, Jupiter
	10°	20°	30°	

The next Triplicity begins with Leo sign, the decanates of which are said to be ruled by Sun, Jupiter, Mars, and so of the rest, the first decan being always ruled by the same planet as is attributed to the sign itself, and the other decans by the rulers of the succedent signs of the same Triplicity.

In the Occidental system, however, Aries is ruled by Mars, and Mars rules the first decan, the Sun rules the second decan, and Venus the third decan. Mercury then follows with the first decan of Taurus, Moon with the second decan, and Saturn with the third decan of that sign. Jupiter rules the first decan of Gemini, Mars the second decan, and Sun the third decan. This order continues through the 36 decans of the zodiac, and ends as it began, with Mars.

From considerable experience of the effects of eclipses I have no hesitation in recommending the use of the Oriental system.

Having determined what planet rules the decan which is occupied by the eclipsed luminary, we have next to find its position and government in the horoscope for the lunation, and this, of course, will differ very widely, according to the longitude and latitude of the place for which the horoscope is set. But for the general import of an eclipse we must give precedence to the nature of the sign in which it falls, considering whether it be in a cardinal, fixed, or mutable sign, and whether it has relation to one or other of the elements. For the cardinal signs have relation to things political, changes of government and rulership, mutations of State affairs, and revolutions among the people. Fixed signs have particular reference to seismic convulsions, the condition of the Earth and its crops or produce, to mines and the interior parts of the

Earth. Common or mutable signs have relation to the condition of the people and their commerce and merchandise. Similarly, the human signs are in a particular sense related to human beings, bestial signs to the condition of cattle, and aquatic signs to the waters of the Earth and the produce of them. Thus it is possible to discern between affairs of State and political affairs, affairs of common or public interest and matters relating to the body of the Earth itself. Also we can refer the effects of eclipses to the human, bestial, or aquatic worlds, and these two sets of considerations will enable us to define with considerable clearness the particular significance attaching to any eclipse of the Sun or Moon.

The eclipse of April 17th, 1912, fell in Aries, a sign ruled by Mars. It was visible over the greater part of western Europe. The sign Aries is a military sign, and this eclipse was soon followed by the Balkan War, which was the precursor of the Great War in Europe.

The eclipse of March, 1914, fell in Virgo 21°, and threatened the crops and food-stuffs of the world, which presently were affected by the outbreak of war and the depletion of the agricultural districts of workers on the land. Food prices increased 50 per cent. in many countries, and clothing, equipment, etc., was in such great demand that supplies gave out, and all sorts of irregularities had to be admitted to the services. There was an eclipse of the Sun in Pisces 6° in February, 1914, and one of the Moon on September 3rd of the same year in Pisces 12°, which threatened the safety, condition, and supply of fish. As a consequence of the mining of the ocean and the activities of submarine craft, not to mention the dangers to trawlers carrying on the fishing trade, the dearth of fish soon became apparent, and a rise in prices amounting to as much as 100 per cent. was soon registered. Pisces is a 12th House sign, corresponding with the 12th division of the visible heavens, and at this time there was a great internment of aliens and rounding up of enemies within the country, and this took place throughout all the belligerent countries: Great Britain, Germany, Russia, France, Austria-Hungary, Servia, Turkey, and Italy. This is agreeable to the general ascription of the 12th House as that of "detention, imprisonment, restraint, internment, ambush, exile, and asylum." Hospitals and places of detention were of course soon filled to their utmost capacity, and vast new quarters had to be found for the hundreds of thousands of wounded.

On August 21st there was an eclipse of the Sun in Leo. This is a 5th House sign, and is ruled by the Sun. It therefore denotes hurt to the uprising generation and great mortality among the youth of the country. It also indicates hurt to kings and rulers, the Sun being the natural significator of those in authority, as the Moon is that of the people.

The eclipses of 1915 fell in the signs of Aquarius and Leo, the Sun being eclipsed on February 14th in Aquarius 25° and in Leo 17° on August 10th. Now the eclipse preceding the Great Plague fell in Aquarius, which, according to medical astrology, is the sign indicating the blood. There accordingly broke out in various parts of the war area a form of typhoid akin to the Black Plague, and throughout Servia and various parts of Hungary the plague was responsible for thousands of deaths. London has already been mentioned in connection with the meridian position of the sign Aquarius, and it is to be noted that the Coalition

Government followed this eclipse, consequent upon a great deal of public censure of the Ministry, which was accordingly reconstructed so as to include an equal number of representatives of both parties. Aquarius as the 11th House sign is related to the Ministry of the country, and Cabinet changes have taken place in most of the European capitals in consequence of the military needs of the various countries.

Thus it is seen how, by mere consideration of the nature of the sign of eclipse, its ruler or planetary symbol, and its correspondence with a particular House or division of the heaven, the import of an eclipse may be determined with some degree of accuracy.

Eclipses of the Sun are especially related to the rulers of nations, kings, princes, presidents, and monarchs of all orders, while eclipses of the Moon are especially related to the affairs and condition of the people.

The following are the various effects symbolised by eclipses of the Sun and Moon according to Ramsey, an old author much quoted.

CHAPTER IX.

The Decanates.

THE following are the indications of eclipses according to the decanates in which they fall according to Ramsey and other old authors.

OF THE SUN.

Aries.—In the first decan of this sign the Sun being eclipsed signifies the frequent motion of armies, continual expeditions of assaults and batteries, with great tumults and seditions and controversies and an inclination of the air to excessive drought.

In the second decan it denotes the imprisonment and sadness of some king and danger of death to him, the corruption of fruit-bearing trees and of things growing upon the earth.

In the third decan it brings grief and sadness to men, the death of a great woman and the destruction of cattle.

Taurus.—In the first decan it destroys compacts and agreements, hurts commerce and business, and brings the crops to nought.

In the second decan it denotes danger to travellers and those that are with child.

In the third decan, pestilence and famine.

Gemini.—In the first decan it denotes dissension, strife and sedition, especially among the clergy of whatsoever order; it stirs up hatreds and feuds and brings neglect and contempt of the laws of God and Man.

In the second it denotes piracies, murders and thefts.

In the third the death of some king, and grave injuries to the interests of the people.

Cancer.—In the first decan it signifies great disturbances in the air and changes of weather. It troubles women and threatens death to a queen.

In the second decan it dries up rivers and fountains and causes outbreaks and wantonness among men and women.

In the third decan it signifies the outbreak of sedition and plagues among the peoples governed by Cancer.

Leo.—In the first decan of this sign it denotes the death of some famous prince and penury of corn.

In the second decan it shows troubles and anxieties to princes and great men.

In the third, captivities, besieging and ransacking of towns, profanation of sacred places, as churches and the like.

Virgo.—In the first decan it argues death or other great calamity to some king.

In the second decan, famine, pestilence and sedition.

In the third, it signifies destruction, banishment, and adversity to painters and penmen and such as live by the brush or pen.

Libra.—In the first decan, it corrupts the air and causes scarcity and dearness of corn.

In the second decan it denotes the death of a king, seditions and famine.

In the third decan, controversies among the nobility and detriment to their estate.

Scorpio.—In the first decan it moves and stirs up wars and slaughter, tumults, hatred, captivity and treason.

In the second, mischief to some king whose mind is averse to war.

In the third, the rise of some tyrant, the idleness and slothfulness of the former king hateful to all.

Sagittarius.—In the first decan it shows dissensions and inveterate hatreds among men.

In the second, the death of camels and other such cattle as chew the cud.

In the third it variously damnifieth horses and armies.

Capricornus.—In the first decan it denotes unhappiness and evil to great men, the transmigration of some king, the rebellion of the nobles and common people.

In the second it stirreth up the sworn or hired soldier against his commander.

In the third it induces to a tumult of activity in the king and causeth famine.

Aquarius.—In the first decan it causes public grief and sorrow.

In the second decan, public robberies, thefts, outrages, earthquakes and famines.

In the third, death and slaughter of sheep and beasts of the field.

Pisces.—In the first decan it dries up rivers and makes the sea-coasts unfortunate.

In the second it shows the death of famous and excellent men, destruction of fish, and causes earthquakes.

In the third it denotes the sedition, fierceness, cruelty and inhumanity of soldiers.

It is obvious that here the author has recourse to the Chaldean order of the decanate rulers, beginning with Mars in the first decan of Aries and continuing on with Sun, Venus, Mercury, Moon, etc., throughout the several successive decans, ending with Mars in the last decan of Pisces. It will also be seen that the various significations have respect to the common astrological rule regarding the kind of things and persons ruled by (a) the sign in which the eclipse takes place, (b) the planet ruling the decan, and (c) the general portent of some disaster affecting these by reason of eclipse influence. It is, moreover, quite apparent that many of the ascriptions are partial and fanciful, and have only a local application, having apparently been written up from the sequence of eclipse happenings during troublous times in this and other European countries. It never fails to bring in the soldiery when Mars is the ruler of the decan, nor to threaten the king with death when the Sun is the ruler, while women under Venus and the Moon fare badly enough according to all the standards of a civilized community. The above remarks apply also to the lunar eclipse effects which follow.

Of the Moon.

Aries.—First decan denotes fevers, house-burnings, firing of woods, and dryness of the air.

Second decan, pestilence.

Third decan, abortive births, incommodities, and such like dangers among women.

Taurus.—First decan, death and disease among cattle.

Second decan, the death of some queen and a dearth of seeds sown and scarcity of produce.

Third decan, serpents and creeping things are affected.

Gemini.—First decan threatens incursions of enemies and rapines.

Second decan, sudden motion of armies and commandeering of private and public things.

Third decan, death of some illustrious man.

Cancer.—First decan, the Moon being here eclipsed denotes the stirring up of wars.

Second decan, grievous exactions, intolerable tributes, taxes, and the like burthens.

Third decan, death to the female sex, sudden destruction and miseries.

Leo.—First decan, sudden infirmity of the king or the death of a great man.

Second decan, journey of the king and mutation of things.

Third decan, stirs up the people and armies to new attempts.

Virgo.—First decan causeth sickness and infirmaties to the king, and various seditions and discords among men.

Second decan, great damage to counsellors and scribes.

Third decan, diseases to mortals.

Libra.—First decan provoketh furious storms of hail.

Second decan is pernicious to everyone.

Third decan threatens the death of some renowned and illustrious man.

Scorpio.—First decan, horrible thunder and lightning, and perhaps an earthquake.

Second decan dries up olives and the air, and causes burning fevers.

Third decan, sharp sickness, with many seditions, quarrels and slaughter.

Sagittarius.—First decan presages thefts and rapines.

Second decan, destruction to horses and mules.

Third decan, the pestilence and many evils among men.

Capricornus.—First decan, instigations among men and the lamentable murder of some excellent man.

Second decan, frequent incursions and assaults of soldiers, robberies and captivities.

Third decan, the death of some king, and also sedition.

Aquarius.—First decan, the sickness of some king.

Second decan universally hurts the seed of the earth.

Third decan causeth a change in all things.

Pisces.—First decan brings sorrow to priests and religious houses.

Second decan, death of some great illustrious person.

Third decan threatens robberies and promiscuous rapines both by land and sea.

Thus far the author, who evidently lived in troublous times. It must be confessed, however, that the ascriptions are exceedingly vague, and in some instances unduly boisterous and violent, while it is difficult to trace the connection between eclipses of the Moon in any sign and violent hailstorms unless they happen immediately at or very soon after the eclipse. But in some particulars it will be found that they have a remarkably close application to the trend of events as we find them.

CHAPTER X.

Transits over Eclipse Points.

THE longitude of an eclipsed luminary being noted, the subsequent transits of the major planets both by conjunction and opposition are valuable pointers as to the exact time at which an eclipse will eventuate, or events symbolised will be realised.

This seems to indicate that the eclipse in itself has no causative value so far as human happenings are concerned, but rather a symbolical signification which point to events which are signalled. It is as if one should say the clock showing seven of the evening indicates the hour for dinner. It cannot be said that the clock either produces, sets out, or serves the dinner, but simply that it indicates the dinner-hour. Similarly a white light showing down the railroad indicates all clear ahead, a green light is a caution as to time, while a red light is a danger signal. It is open to the driver of a locomotive upon seeing the red light either to slow down and stop until the signals are favourable to progess, or to go ahead with full steam and meet with disaster. Eclipses have no compelling influence; they are not efficient causes in a moral and intellectual sense, though they may very well have a direct dynamic effect upon the body of the Earth itself, existing and operating as they do upon the same plane. Also they may have responsive action in the physical body of man, it being compounded of cosmic elements similar to, if not identical with, those that enter into the composition of all celestial bodies. In this sense the Sun is related to the organic nature of the body of man, and the Moon to the functional nature. But of this more anon.

The significance of an eclipse having been determined by its sign position, it will be found that what is signified usually comes to pass exactly at the time that one of the major planets afterwards makes transit of the longitude of the eclipse, whether by conjunction or opposition.

Thus the eclipse of April 17th, 1912, which was the precursor of the Balkan War, took place in Aries 27°. Now the planet Mars came to the opposition of that eclipse, that is to say the longitude of the Sun, on October 14th, 1912, and on that very day war broke out at Adrianople. It is worthy of note that the eclipse line of apparition fell in latitude 43° North, and its longitude was 27° East of the equinox, which, measured from Greenwich, points exactly to the centre of outbreak, the Balkans. The war took its course, and came to an apparent end, but suddenly broke out again in June, 1913, on the very day that Mars came to the place of the same eclipse in Aries 27°. From this we learn that eclipses, or the events signified by them, may remain in latency for many months, and are liable to spring into existence at such times as the major planets shall transit them. It further informs us that although certain physical effects due to the dynamic action of the luminaries on the earth, such as tidal waves, earthquakes, and atmospheric phenomena, may take place immediately or very soon after the eclipse, yet there are other events which can only have a human interpretation, and these are indicated by the transits of the planets—

within certain limits of time from the date of the eclipse.

This time-measure of eclipses needs some consideration. It has to be known how long an eclipse can be said to be effective. I use the term for want of a better.

The ancient writers have given a concise rule, which shows that the duration of an eclipse influence is to be measured by the number of hours the luminary is under eclipse from the first to the last contact. If the Sun is eclipsed they will indicate so many years, but if the Moon, then so many months. Thus if the time from first to last contact is two hours in the case of a solar eclipse, then the influence of that eclipse will remain over the world for two years, and the transits over the place of the eclipse and its opposition will show the times when crises will take place. And in the case of a lunar eclipse, if the duration from the Moon's first contact with the shadow to its last contact be six hours, then the effects will last for six months. This observation clearly shows that the ancients were aware that eclipse effects do not happen immediately, but that there is a process of exhaustion, which takes time.

I have already pointed to the lunar eclipse of March 12th, 1914, as the Great War eclipse. This fell in Virgo 21°. Its particular pointing I shall deal with in another section. What I would point out here is that this eclipse produced apparently no active results until Mars (here, as in every case, the special indicator of war and strife) passed over the 21st degree of the sign Virgo, which happened on July 30th, 1914, at which time Germany, Russia, and France were suddenly plunged into conflict, Great Britain joining issue in defence of Belgian neutrality, and as an ally of France and Russia, on August 3rd, by declaration of war against Germany. This may be regarded as a coincidence, but if so it should not be forgotten that coincidencies make laws. There are exceptions to all laws, and reasons for these exceptions, but it is from the coincidence of phenomena that we are able to formulate a law. There are exceptions, for instance, to the tidal law, and although it is the normal fact that high tide occurs twice in a day, yet there are places where there is only one daily tide and others where there is no tide at all. When a prediction, based on these astronomical factors, turns out to be correct, it is usually referred to by the ignorant as a "coincidence," by which I presume is meant a coincidence of the event with the prediction, for there is no other coincidence to be noted, unless it be that similar indications are followed by similar events, as in the cases cited above.

The nature of the planet which transits the place of the eclipse would rationally appear to have much to do with the kind of results that are indicated, as Mars for wars and strife, fires, outbursts of popular feeling, etc.; Saturn for mortality and dearth, famines and privations of all kinds; Uranus for revolts and insurrections, strikes, and discords and sudden catastrophies causing fractures and dislocations among the people; Neptune for ambushes, secret assaults, treachery and covert malice and intrigue; while Jupiter, through his expansive power, may indicate earthquakes and violet storms, magnificence and extravagance among the rulers and people.

Also it should be observed what planets are in conjunction, opposition or quadrature to the eclipse, for these will serve to indicate the causes which lead to the events signified. Thus in the case of the eclipse which happened on 27th July, 1664, being a total eclipse of the Moon in Aquarius 14 we find Uranus in conjunction with the Moon in the sign Aquarius which indicated that mysterious and extraordinary affection of the blood which attended the Plague, and moreover the dislocations and separations which afflicted the people in consequence of their fleeing from the area of infection. The eclipse took place at 11.2 p.m., when the Sun was in the 4th angle and the Moon in the 10th, Mars in the 6th House in Libra, and Saturn exactly on the cusp of the 8th House. These are indications of a raging fever or pestilence and much mortality.

The duration of eclipses has already been indicated as depending on the hours of obscuration, which show as many years in the case of a solar eclipse and as many months in the case of a lunar eclipse.

But as to the time at which these effects begin to appear we have the following rule by Ptolemy (Quad. 1, 6.): If the eclipse fall in the Eastern horizon, the effects thereof will first manifest themselves about the next four months following the eclipse, and will most strongly appear about the first third part of its whole duration.

If it fall in the Midheaven the events thereof will begin to appear in the fourth month next following, but most apparent will the effects be in the middlemost third part of the whole duration. If it happen in the West part of the horizon, the effects shall not begin until about the last four months, and its greatest operation will be in the last third part of its whole continuance.

Here it is seen that Ptolemy makes the whole duration to be one year, of which the East controls the first four months, the Midheaven the middle four months, and the West the last four months, these being the times when the effects of an eclipse are most apparent according to the quarter, East, South or West, in which it may occur. But nothing is said of eclipses that happen in the lower meridian, which may be as remarkable and effectual as any others, seeing that eclipses are symbols and not causes in a human sense, whatever we may say of their action on the atmosphere and the body of the earth itself. Doubtless the maxim: What the eye does not see the heart does not grieve, has had some bearing on the general idea that eclipses must be visible to be effectual, but this I consider to be subversive of the entire system of astrology, inasmuch as we could thence argue that planets under the horizon have no significance in the life at all, whereas it has been continually our experience that, in their several departments and according to their respective natures, they are as significant as others that are above the horizon; and in the belief and judgment of some modern writers, planets in the lower angle are as potent as those in the Midheaven.

Probably the matter could be equated by taking the duration as one year and allowing three months for each of the angles in succession, namely, the East for the first three, the Midheaven for the second three, the West for the third period of three months, and the North for the last three. Then, having the duration of the eclipse according to the dimensions of it as above stated, we can make proportion accordingly. Thus if a solar eclipse were to last one and a half hours from first to last contact, then each of the four quarters would have four and a half months allowed to them, and if the eclipse fell in the South, the effects signified would be observable in the period from 4½ months after the eclipse to 9 months after it, that is to say, in the second of such periods.

But there is a rule which has reference to both sides of the horizon, and from which the time of the beginning of the effects may be more accurately determined. It requires that the distance of the eclipsed luminary from the horizon it last crossed, namely, the East for its rising or the West for its setting, shall be taken into account and a proportion made with the whole time it is above or below the horizon, which period is equal to one year. Thus if the eclipse be one of the Sun and he were eclipsed at four hours after his rising, his whole arc being at that season of the year 8h., then the eclipse would begin to act at six months from the time of its occurrence. In the case of an eclipse of the Moon we have to take the time from the last rising or setting until the middle of the eclipse and make the same proportion. Thus if the Moon rose at 6h. 17m. p.m. and the eclipse took place at 11h. 24m. p.m., then the period from the rising to the middle would be 5h. 7m., and if the whole arc of the Moon, or the time it remained above the Earth—which can be found in the almanac—be 14h. 5m., the proportion would be :

As 14h. 5m. is to 365 days, so is 5h. 7m. to 133 days,

which, added to the date of the eclipse, will give the date at which the effects will begin to appear; and they will last for as many years or months as there were hours of obscuration, according to whether it was a solar or lunar eclipse.

So much then in regard to the old authors, and it would be easy to disprove their conclusions while admiring their ingenuity. I will cite only two instances: 1. The eclipse of June 6th, 1853, fell on the Midheaven at Panama, and within six weeks there followed a terrible earthquake in which 4,000 people were killed and immense damage done to property. Now, as this was a complete solar one, the duration from first to last contact would be about two hours, and this would indicate two years for the duration of the effects. In the locality named the day would be 6h. 20m., and as the Sun was then in the Midheaven, the half of its course would be completed since sunrise. Hence the effects should begin, according to Ptolemy, only after eight months had elapsed, and according to other authors only after three months. But we see that they were realised in rather less than half the time, namely, in less than six weeks.

2. The eclipse of May 7th, 1902. There was an eclipse of the Sun which passed directly over the Antilles, and on the very day of the eclipse there was an eruption of Mont Pelée, in the island of Martinique, which devastated that place completely. Here we find no measure of time, but that which is counted from the meridian.

It is therefore obvious that no correct rule has yet been given as to the time at which the effects of eclipses may be expected to commence. I have found from considerable study that remarkable effects do most assuredly happen at the times when the major planets transit the place or opposition of a preceding eclipse, providing that such transit happens during the period included in the limits of time ascribed to its continuation as determined by the hours of obscuration. But whether these are the first effects which could be legitimately ascribed to such eclipse is a matter which involves rather more knowledge of history than I pretend to. This, however, is the most dependable rule that I can offer at the present stage, and that it is a sound and uniformly consistent rule I am well assured.

In the case of the Panama eclipse, for instance, the Sun was totally eclipsed on 6th June in ♊ 16°, having N. declination 22° 43′, which answers to the latitude of the West Indies. The eclipse would be on the meridian 8h. West and would affect the whole of the Central American earthquake area. The point, however, is that on 15th July, 1853, we find the planet Jupiter in Sagittarius 16° and Mars in Gemini 16° directly affecting the line of eclipse, while five planets were in Fixed Signs. It would be difficult to find a more luminous illustration of my rule; yet it is by no means an isolated case.

CHAPTER XI.

Individuals and Eclipses.

When we come to consider the "effects" of eclipses, that is to say, those events in human life of which it may be said the eclipses are portents or signals, we have to admit that these events could not transpire but by the mediation of certain individuals whose minds are so set as to render them capable of and even disposed to the realization of such social and political conditions as are signified. For thus the universe is governed, not immediately, but by mediation or agency from the Archangel down to the lowest and most servile of the human race, the bestial world, and beyond this, unconscious agents—if consciousness can ever be dissociated from matter—such as the elements. For if we are rightly to predict such things as wars, tumults and oppressions, we rationally infer the existence and co-operation of human wills disposed to the stirring up of strife and of others defending the cause of peace. Also, we cannot predict plagues and pestilences without inferring also a corrupt state of the blood and the existence of certain germs which are more than usually prevalent at a certain time, and thus we argue an interaction of the inorganic and organic worlds, and also, perhaps, a state of mind which engenders a condition of health in man which predisposes him to contract and engender disease. It is, indeed, a complex philosophical problem as to what interpretation we are entitled to put upon the facts of astrological science, but if we are rightly disposed in regarding eclipses as portents, then it is forced in upon us that the only means by which the events signified can come to pass is by mediation, that is to say, by the voluntary co-operation of human agency. And having regard to the nature of some such events, it will appear altogether blasphemous to the religious mind to ascribe these things to the Will of God. Rather we should say certain things happen by the Divine Will and others by Divine Permission. For, in order that man's conscious evolution can be progressively realized, a certain degree of freedom must be allowed to him. But that his will can be ultimately effectual against the Divine Will is not consistent with our views of an intelligible universe governed by a Supreme Intelligence. Thus man may hinder, but he cannot frustrate the purpose of life. His hindrance can only be local, and not universal. Hence we are not in conflict with religious thought in regarding eclipses as portents. Rather are we illustrating the operations of a divine Goodness and fatherly Providence in suggesting that there never is any great calamity or hurt befalling the world but it has been duly signalled, and we ourselves thereby warned to dispose ourselves to its remedy. Thus we admit that every man can do his own will to the extent of his individual power, and in so doing he is either in harmony with or contrary to the pre-established order of things which we connote with universal law, and thus he also prospers or suffers in consequence, and so gains experience of what we call good and evil in the world, which experience serves the higher purposes of the Soul. For if the world were entirely abandoned to the wills of men, and there were no general direction of the course and issue of events, the earth would be speedily plunged into chaos, and so cut off from the mediation of the superior worlds that it would suffer extinction. And this regulation of the course of events, happening as it does according to universal laws, is determined by the courses of the stars in regard to any particular planet, so that certain cosmic changes are continually going on in respect to it, and from hour to hour a new die is struck, as shown by the relative positions of the celestial bodies both among themselves and in relation to a particular locality, which we call the horoscope, and which answers to a particular note, vibration, or form of energy to which a set of individual Souls respond, and these accordingly take birth and become human agents. Thus the Supreme Will is carried out by mediation, and thus is proved true that old Chinese maxim which says: Heaven does the work and men are only its agents.

Now, inasmuch as we have seen that eclipses are portents of things in general, they must also be portents of things in particular, and this follows from the fact that collective results are individually conspired to, that a mass consists of its units, that humanity is composed of individuals, a kingdom of its subjects, a community of its members. So that we are only reasonable in regarding eclipses as having an individual as well as a collective significance. It is in relation to individuals that we have now to consider eclipses.

It should first be stated that there are four chief points in a horoscope which are regarded as Significators. These are the Midheaven, the Ascendant, the Sun and the Moon. The positions of these by longitude in the horoscope of birth should be noted, and it will invariably be found that a particular and permanent signification attaches to these positions throughout the life of an individual. For if an eclipse falls in conjunction with or opposition to any one of these points, there will follow a period of sickness, stress, misfortune and disfavour which cannot fail to arrest the attention of the most casual student.

It should further be observed that the Sun is indicative of the father and the male portion of the family circle, while the Moon denotes the mother and the female side of the family. Also the Sun denotes organic diseases, while the Moon denotes functional disorders. Thus the Sun stands for the vitality, while the Moon is the purveyor of this vitality to the several parts of the organism.

Moreover, the Sun has affinity with the Midheaven and the Moon with the Ascendant. Thus the Sun and Midheaven have signification of the position, honour, credit, profession and status of the individual, while the Moon and the Ascendant have relation to his health, general affairs, public relationships, and the changes which take place in these.

Also the Midheaven denotes the position to which he aspires, his ambitions and hopes, the Ascendant to that which he gives, the things he does, and his output of energy, the West angle or Descendant to that which he suffers, what he receives or that which is done to him by the action of others, and the Nadir or 4th angle has relation to his locality, home, estate, and that which he stands upon or holds in tenure or possession. These are natural correspondences, as may be seen from the following diagram of the horoscopical man.

In judging the effects of an eclipse we have therefore to observe in what quarter of the heavens it may fall as regards his own horoscope, what Significator or planet it shall coincide with, and what House or division of the heavens it may be posited in at the time of its centrality.

For it is to be observed that all the planets, as well

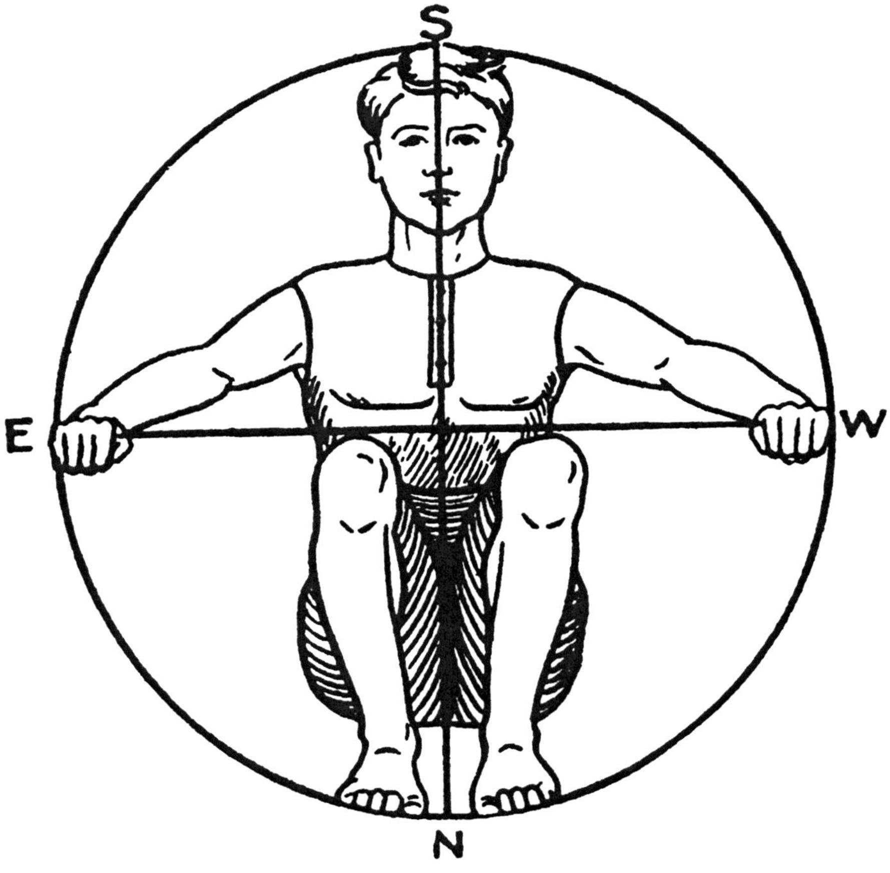

Fig 7.

as the four Significators above mentioned, are indicative of certain things and relationships, as Mars of brothers and collateral male kindred, Venus of sisters and collateral female kindred, Jupiter of uncles, Mercury of aunts, Saturn of aged persons and old associations, and Uranus of grandparents.

Moreover, Saturn denotes losses and privations, ill-health and disease. Uranus denotes dislocations, removals, separations, sundering of ties and violent dissensions. Mars denotes fevers and accidents and dangers by fire and arms, sometimes theft and violence, scalds, burns, wounds of various sorts, according to the sign it occupies, and also frenzies and madness. Venus denotes social and domestic affairs and concerns of love, marriage, etc. Jupiter causes congestions, surfeits, extravagance, and losses through undue optimism and over-confidence. Mercury denotes commerce and business affairs, details of the life, which by eclipse cause worries and anxieties and sometimes dangers by travelling.

The 1st House shows the person, the 2nd finance, the 3rd journeys and relations, the 4th the house or residence, the 5th speculation, pleasures and domestic affairs, the 6th servants and personal comforts, food, clothing, etc., the 7th contracts, agreements, treaties, rivals and things that are counter to the person, the 8th death, wills, legacies, the wife or other partner's finance, the 9th foreign affairs, religious concerns, the 10th credit and position, superiors, the 11th associates and advisers, friends and colleagues, the 12th the sick-room, hospital, restraints, confinements, sequestration and exile.

When hurts are signified they should be judged by the sign in which the significant planet may be, that is, the planet on which or in opposition to which the eclipse may fall, as to whether they be fiery, airy, watery or earth, whether they be bestial, human or reptilian. Thus the effects of eclipses may be judged with considerable accuracy. These rules are agreeable to the dictum of Ptolemy, who says that an eclipse will operate most efficaciously . . . "upon such people whose nativities agree with the eclipse, that is to say, upon them whose Nativity or Revolution have the place of the horoscope, Sun, Moon, part of fortune, and other planets, in or about the degree of the eclipse or the quadrangle thereof."

An eclipse must always be considered as a privation or misfortune of some sort, inasmuch as the cutting off of the light of the Sun or the reflected light of the Moon has this general significance, from the correspondence that darkness has with things that are baneful and evil, and whereas the eclipse of the Sun is more significant of hurt to the king, the father, the head male representative of the family, etc., so an eclipse of the Moon has more significance of the queen, of the mother, the head female representative and her dependents. And from this institution of Nature it appears that there is no effective revolt, seeing that the Sun universally denotes the male and the Moon the female, each being necessary in its particular function to the upholding of our terrestrial integrity and welfare, their conjunctions being the period of natural generation, their oppositions the times of their bringing forth, and their eclipses those of their mutual afflictions. For each is necessary to the other and both are regnant in their own particular spheres.

In observing the position of an eclipse it should be noticed whether the planet or Significator on which the eclipse falls in the horoscope is harmoniously or inharmoniously configurated, whether it be upheld by trines and sextiles of other bodies or afflicted by squares and oppositions, for these will determine the degree to which an eclipse is capable of affecting the condition of things, on the principle that a man who is weak and of poor standing can more readily be assailed by misfortune than one who is strong and affluent. Thus, if the Midheaven at the birth be afflicted by the quadrature of Saturn, and an eclipse falls on the Midheaven, the honour and credit, position and standing of the subject will be readily undermined inasmuch as it is radically weakened by the quadrature of Saturn, whose radical import of evil is thus brought into effect by the eclipse, the latter being the time-signal which warns of this calamity. And this is something that is desirable to know, even though it may be found that the sagacity of man is insufficient to wholly thwart and intercept the threatened evil, for we have at least the satisfaction of knowing that these things, without let or hindrance, are governed by laws that are intelligible and within the benevolent control of that Intelligence of which they are the natural expression.

It has to be remembered when considering the value of eclipses as portents that the ancients had nothing but these and the transits of the planets to guide them in their study of astrology, and that by means of them they were able to trace a definite connection between the changes taking place in the heavens and the course of human events, so that by an anticipation of celestial motions they could predict events with singular definition and clearness, of which there are many illustrations both in sacred and profane writings.

CHAPTER XII.

Illustrations.

The following illustrations will doubtless be appreciated by students in so far as they tend to show that there is a certain coincidence of events with eclipses and the means by which from like causes we may predict like effects.

B.C. 721, March 19th, a total eclipse of the Moon in the sign Virgo. Babylon was immediately afterwards captured by Sargon. [Reference to the government of the countries by the various Signs of Zodiac, as given by Ptolemy, will show that Babylon was ruled by Virgo.]

B.C. 585, May 28th, an eclipse of the Sun, foretold by Thales of Miletus, by which the Lydians capitulated to the Medes and brought an end to the war [for " it happened that as the battle was at its heat the Sun was suddenly darkened and the day turned into night." Johnson links this eclipse with the prophecy of Isaiah (xiii.-10) about the overthrow of the Babylonians. It happened in Gemini, the meridian sign of Babylon, as Virgo was its ascendant.]

B.C. 523, July 16th, an eclipse of the Moon, followed by the death of Cambyses.

B.C. 502, November 19th, an eclipse of the Moon, followed by the slaughter of the Sabines and the death of Valerius Publicola.

B.C. 460, April 30, an eclipse of the Sun, followed by the Persian War and the recession of the Persians from Egypt.

B.C. 431, April 25th, an eclipse of the Moon, and August 3rd, a total eclipse of the Sun; these were followed by a great famine in Rome, the plague at Athens, and the outbreak of the Peloponnesian War. [This war lasted for 29 years, and it therefore shows that there is no reliance to be put upon the rule for the duration of the effects according to the hours of eclipse, for no eclipse of the Sun ever lasted for 29 hours, or more than a twelfth part of it. But it is of interest to note that it took place in the sign Leo, which, according to all the old authors, is the ruling sign of the City of Rome, its meridian sign being Taurus, *i.e.*, Italos, the Bull.]

B.C. 413, August 27th, a total eclipse of the Moon. Nicias and his ship destroyed at Syracuse. [This eclipse would fall in the sign Pisces, a watery sign, and the eclipse of the Sun would fall in Virgo that year, in the constellation Argos, the ship.]

B.C. 394, August 14th, an eclipse of the Sun. The Persians were beaten by Conon in a marine engagement.

B.C. 168, June 21st, a total eclipse of the Moon. The following day Perseus, King of Macedonia, was defeated by Panlus Emilius.

These eclipses are all from the record of Ricciolus, whose remarks as to their fulfilment are of interest, and to these I have added some comments of my own, the same being included within brackets.

But to come nearer to our own time, we have already cited the total lunar eclipse of A.D. 1664, which fell in the second decan of Aquarius, a human sign, in the meridian of London and near to the Mid-heaven of the City at the time of its occurrence, which was followed by the Plague. There was another of the Moon on January 30th, 1665, which fell in Leo, close to the place of Saturn, in the lower angle of the City of London radix, and which was followed by the Great Fire, the sign Leo being a fiery sign. There was also an eclipse of the Sun on July 12th, in the same year, which fell near the place of Mars in that radix.

Then there are the eclipses at Panama and Martinique, which were immediately followed by great earthquakes in those parts, and which luminously confirm the dictum of Ptolemy that " they operate most efficaciously in those places where they are vertical, or where the chief Significator shall pass by their zenith in the time of the eclipse (lib. 2, cap. 8)." By the chief Significator I understand him to mean that luminary which suffers eclipse. This agrees in part with the dictum of Cardan, who says that " eclipses operate most particularly in those parts that are governed by the sign in which the eclipse takes place, and those also where it is visible." Obviously, therefore, when we find the eclipse taking place in any locality, and that place is also ruled by the sign of eclipse, the effects would be most assuredly felt in that part of the world. But an eclipse need not be visible in a country in order to produce most disastrous effects therein, for if it happen in the ruling sign of that country, whether it be the meridian sign or the sign ascending, it will still bring about such disasters as are signified, that is to say, it will signal or portend them. Similarly it is sufficient that an eclipse is vertical without being in the ruling sign in order to produce marked effects. But this will appear on examination, namely, that when the eclipse falls in the meridian sign of a country or city the Government and rulers of that place are effected, whereas if it falls in the ascending sign, the people are principally affected. Also when it is visible in the zenith of a locality that part of the world suffers a series of physical ills, as earthquakes, storms, tidal waves, eruptions, and other evils.

But we may now turn our attention to the significance of eclipses in individual horoscopes, these being of more particular interest to students of astrology, beside being more easily verifiable by reason of the closer scrutiny that we are disposed to give them.

In May, 1902, there was eclipse of the Sun which fell in Taurus 17°, in direct opposition to the place of the Sun in King Edward VII.'s horoscope (November 9th, 1841), and the Sun at his birth was afflicted by the semisquare aspect of Saturn and the semisquare aspect of the Moon, as well as the square aspect of Neptune. It was, therefore, a vulnerable point. The king was to have been crowned at the end of June, but from consideration of this portent I was able to say that His Majesty would not complete his intention, as at that time he would be suddenly struck down by an illness, which I verily thought to be mortal, incidental to the excretory system. This I caused to be known to many students of astrology prior to the event. As events turned out the king was suddenly taken ill on the eve of his projected coronation by a severe attack of appendicitis, which required immediate operation. It was not until August following that the ceremony took place.

Prior to this event there had been an eclipse of the Sun, total, in the last degree of Scorpio, which took place on the 22nd November, 1900. This was close to the opposition of the Ascendant of Queen Victoria's horoscope, the Node being exactly on that point of the zodiac. Within two months the aged, venerable and much beloved sovereign passed away, death taking place on the 22nd January, 1901.

The death of King Edward, so clearly foreseen and predicted by me in the current almanacs of the year 1910, was duly signified by the large eclipse of the Sun which took place in Gemini 26°, in close opposition to the Ascendant of the royal horoscope and near the opposition of Saturn, which was rising in the horoscope in square aspect to the Moon. The king died on the 6th May, 1910, within two or three days of a solar eclipse in exact opposition to the place of the Sun in his horoscope, the eclipse taking place in Taurus 17 and the Sun at his birth being in Scorpio 17.

The same event was duly signified in the Prince of Wales' horoscope, for on the 4th June, 1909, there was a total eclipse of the Moon in Sagittarius 13, which fell in direct opposition to the place of the Sun in the radical horoscope of the Prince, the Sun being here the natural Significator of the father, as already laid down in the rules concerning the interpretation of eclipse influences.

On the 26th September, 1912, the Moon suffered eclipse in Aries 3°, the Moon at that time being close to the lower meridian of the horoscope for London. Very serious indeed were the political and constitutional changes that took place immediately after this eclipse, and when in the following year the Sun was eclipsed in Libra 6, close to the opposition of the radical Ascendant, following upon the eclipse of the Sun, visible in England, on the 17th April, 1912, in direct opposition to the place of Saturn in the King's horoscope, there was undoubted danger of war. For Saturn was at birth in the 7th House, and two eclipses thus happening in the 7th House disturbed the validity and integrity of treaties and agreements existing between the Crown and the rival Power, so that within a year of the last event the Great War broke out, and was very clearly foreseen and predicted in the current almanac under the months of July and August, 1914, the exact date of Germany's declaration of war with France and that of our declaration against Germany being given as dates of international crisis, wholesale expenditure on munitions of war, intense public excitement and depression on the Stock Exchange.

Now the use of the "Ready Reckoner" will inform us that the Node is passing through the sign Aquarius in 1915 and reaches 9° 43′ of that sign on the 1st January, 1916, its course thereafter being through the first degrees of Aquarius until by retrograde motion it reaches Capricorn 20° 21′ on the 1st January, 1917. Inspection of the Table of Lunations on page 36 of that work shows a lunation in Aquarius 13 which will be within 18° 36′ of the Node, and therefore will constitute an eclipse of the Sun. Also another which falls in Leo 6, and which again will be a solar eclipse. Also by the addition of 15 degrees to the positions of the lunations there given, it will be seen that there is a partial lunar eclipse at the end of Capricorn in January.

Now it happens that the planet Mars in the royal horoscope was in Leo 5° in the fifth division of the heavens, and therefore this eclipse of the Sun in Leo 6 is extremely inimical to the firstborn of the progeny, from which we see that there are sudden alarms and dangers attaching to the Prince, and as in the month of January, 1917, the Moon is totally eclipsed in the sign Capricorn and near the place of the progressed Sun in the Prince's horoscope, now in square aspect to Saturn, there are manifest reasons for suspecting some calamity.

There is a total lunar eclipse in January, 1917, which falls close to the Ascendant of the horoscope of Kaiser Wilhelm II., and it is followed in August, 1917, by a transit of Mars over the place of the eclipse and a transit of Saturn in opposition to the place of the Sun at his birth (January 27th, 1859), which signifies the end of the Hohenzollerns.

The eclipse of the Moon in March, 1914, which fell in Virgo 21°, was in direct opposition to the Midheaven in the Kaiser's horoscope, and this signifies his downfall and the loss or eclipse of the whole of his tenure and possessions. Also that solar eclipse of 1916, which falls in Leo 6, is close to the opposition of the radical Sun in the 8th House of the horoscope.

The eclipse of the Sun, August 21st, 1914, fell at the end of Leo, close to the place of the Sun, Moon and Saturn in the 11th House of the horoscope of Franz Josef, Emperor of Austria, and clearly indicated the ruin of his hopes and the great evil done to him by his treacherous ally, who first deceived and afterwards exploited him. It was a clear case of "God save me from my friends!" This eclipse presignified the end of the House of Hapsburg.

It would hardly be consistent with our views of the scientific precision of astrology to adventure further in regard to the future without adequate data, and doubtless sufficient has been said to indicate the course that is followed in judging of the effects of eclipses. If a list of the various eclipses during the course of an individual's life is taken and compared with the various positions of the Significators—the Midheaven, Ascendant, Sun and Moon, in the horoscope—and also with the places of the planets, it will be abundantly shown, not only that they indicate periods immediately following the eclipses which have proved unfavourable to the fortunes or health, but also that the nature of the planet involved has been indicative of the nature of the affliction, taken in relation to the sign in which that planet is found, and also the house in which it falls in the horoscope of birth, and this whenever an eclipse has coincided with one or other of such radical places. Instances might be multiplied indefinitely, but the insistent repetition of these indications can only be adequately appreciated by one who is continually occupied with the examination of the horoscopes of individuals.

CHAPTER XIII.

Conclusion.

I have now given a survey of the theory of eclipses both from the astronomical and astrological points of view, the one depending on the other, since without correct astronomical precepts we can make no clear astrological judgment. In the preparation of this work I have sought chiefly to be of service to the cause of a scientific astrology, believing that whatever progress we may make in astronomical science, such knowledge can only be rendered of practical utility and of benefit to mankind by its application to daily life. Astrology is, so to speak, a first knocking at the door of the Inscrutable. We have the highest authority for the belief that by persistent knocking the door will be opened to us, and when we consider the great heights to which other sciences have attained by dint of unremitting research, tabulation and experiment, we have every reason to believe that the same degree of patient study accorded to the subject of planetary interaction and cosmic symbolism would have results not less satisfactory. It is rather in the direction of a utilitarian and humanized astronomy that I should look for the highest reward of such labours. But Urania is a coy mistress, and we must needs woo her ardently and persistently if we would obtain the favour of her good graces.

For many centuries between the periods of Claudius Ptolemy and Kepler, very little progress, if any at all, was made in the science of astronomy, and it has only recently transpired, through the labours of Professor Newcomb, that our astronomers have an accurate set of astronomical Tables by which to calculate eclipses.

Buckhardt's Tables were used for a considerable time prior to the advent of Newcomb, but it is problematic whether even now the centennial variations of the Moon and planets are altogether accurately known and applied. Buckhardt uses some thirty-seven equations for computing the true place of the Moon in the zodiac, including among these disturbances alleged to be due to the action of Venus and Jupiter upon her body or upon that of the Earth, which is relatively the same thing, but the action of Mars, Saturn and the more distant bodies is left out of consideration. Yet on Newton's theory of the solidarity of the system we should naturally expect that all the planets, and not a few of them, interact, and we have therefore the alternative of including all the planets in our equation of the Moon's motion, or of proceeding from a radical value in which all the interactions are already embodied, and this seems to be the rational viewpoint. For the Moon's motion is after all what it is by reason of such interactions, which results in what we call the mean orbit, and I have discovered that using such orbit as the basis of calculation, there is no other equation required at all than that which is due to the action of the one gravitating centre of our system, the Sun. Consequently I am able to find the Moon's place in the zodiac with five equations, as easily as is now done with thirty-seven. And for periods of more than a century the centennial variation equation has to be applied owing to the action of the Sun upon the Moon, by which it insensibly increases its orbit from year to year, but which in the course of a century is an appreciable increment.

In the course of his work on "The Story of Eclipses," Mr. George F. Chambers, F.R.A.S., has an interesting note in reference to the connection that is believed to exist between eclipses and earthquakes. He says:

"Perhaps this may be a convenient place to make a note of what seems to be a fact, partly established at any rate, even if not wholly established, namely, that there seems some connection between eclipses of the Sun and earthquakes. A German physicist named Ginzel has found a score of coincidences between solar-eclipses and earthquakes in California in the years between 1850 and 1888 inclusive. Of course there were eclipses without earthquakes and earthquakes without eclipses, but twenty coincidences in thirty-eight years seems suggestive of something."

That is the whole point, we have to determine scientifically what this "something" is. The researches of Prof. Sir Norman Lockyer have shown that there is a definite connection between sunspot periods and high Nile tides, and I have elsewhere shown that the period of maximum frequency of sunspots exactly coincides with the periodic times of the planets Mercury, Venus, Earth and Mars, taken together, so that here we are coming to the point where we can rationally argue a connection between the planets and mundane events. But we have no reason to give the palm to the German physicist, seeing that the same observation with more particularity and precision had previously been made by Commander Morrison, R.N., the well-known writer of "Zadkiel's Almanac" and a thoroughly versed student of physics and astronomy, who extended the observations of Ricciolus and laid down definite rules for the prediction of earthquakes, many of which he himself successfully foretold both as to time and place of occurrence.

Had Ginzel extended his observations beyond California to all parts of the world, had he included eclipses of the Moon instead of relying wholly upon those of the Sun, and had he also taken account of the subsequent transit of such eclipse places in the zodiac by the major planets, he would have considerably amplified his list. But as I have already said, many coincidences make a law, and the law of eclipses in connection with earthquakes deserves more attention at the hands of scientific men than it has yet received. The observations here made do not comprehend more than a small portion of those which, during the past thirty years, I have been able to collect, but at the present time I have in hand a work of some dimensions in which the eclipses directly connected with earthquakes for some centuries past have been tabulated so that future research along these lines will, it is hoped, be thereby greatly facilitated. The calculations of eclipses herein given are not presumed to be more than close approximations, within a minute or so of the truth, but this would not materially affect the elements involved, and so they serve my purpose very well. With these remarks I may bring this little work to a close, in the hope that it will have fulfilled the purpose expressed in the preface, and thereby constitute a useful contribution to the category of applied sciences.

Eclipses

Mean New Moons in March

Epoch	March			Sun-Node			Moon's Anomaly			Sun's Anomaly		
	d.	h.	m.	s.	°	'	s.	°	'	s.	°	'
1800	13	0	19	11	3	54	10	7	58	8	23	13
1801	2	9	8	11	11	57	8	17	46	8	12	29
1802	21	6	40	0	20	40	7	23	23	9	0	52
1803	10	15	29	0	28	43	6	3	11	8	20	7
1804	28	13	1	2	7	26	5	8	48	9	8	30
1805	17	21	50	2	15	28	3	18	37	8	27	45
1806	7	6	39	2	23	31	1	28	25	8	17	1
1807	26	4	11	4	2	14	1	4	2	9	5	23
1808	14	13	0	4	10	17	11	13	50	8	24	39
1809	3	21	48	4	18	20	9	23	38	8	13	55
1810	22	19	21	5	27	3	8	29	15	9	2	17
1811	12	4	10	6	5	6	7	9	4	8	21	33
1812	0	12	58	6	13	8	5	18	52	8	10	49
1813	19	10	31	7	21	52	4	24	29	8	29	11
1814	8	19	19	7	29	54	3	4	17	8	18	27
1815	27	16	52	9	8	37	2	9	54	9	6	49
1816	16	1	41	9	16	40	0	19	42	8	26	5
1817	5	10	29	9	24	43	10	29	31	8	15	21
1818	24	8	2	11	3	26	10	5	8	9	3	43
1819	13	16	50	11	11	29	8	14	56	8	22	59
Values for Additional Years												
20	18	14	4	1	16	18	9	12	35	0	18	8
40	7	15	23	2	1	55	6	29	22	0	7	9
60	26	5	27	3	18	13	3	11	57	0	25	17
80	15	6	47	4	3	50	11	28	43	0	14	18
100	4	8	7	4	19	28	8	15	30	0	3	19
200	8	16	33	9	8	56	5	1	0	0	6	38
300	13	0	20	1	28	23	1	16	29	0	9	57
400	17	8	27	6	17	51	10	1	59	0	13	16
500	21	16	33	11	7	19	6	17	29	0	16	36
600	26	0	40	3	26	47	3	2	59	0	19	55
700	0	20	3	7	15	34	10	22	40	11	24	7
800	5	4	9	0	5	2	7	8	10	11	27	27
900	9	12	16	4	24	30	3	23	39	0	0	46
1000	13	20	23	9	13	58	0	9	9	0	4	5

TWO PRINCIPAL EQUATIONS OF MEAN LUNATION.

	1. Due to Sun's Anomaly.						2. Due to Moon's Anomaly.						
	s. —0	s. —1	s. —2	s. —3	s. —4	s. —5	s. +0	s. +1	s. +2	s. +3	s. +4	s. +5	
	h. m.	h. m.	h. m.	h. m.	h. m.	h. m.	h. m.	h. m.	h. m.	h. m.	h. m.	h. m.	
0	0 0	2 3	3 34	4 10	3 39	2 7	0 0	5 13	8 47	9 47	8 9	4 35	30
1	0 4	2 7	3 37	4 10	3 37	2 4	0 11	5 22	8 52	9 45	8 3	4 26	29
2	0 9	2 10	3 39	4 10	3 35	2 0	0 22	5 31	8 56	9 45	7 57	4 17	28
3	0 13	2 14	3 41	4 10	3 32	1 56	0 33	5 40	9 0	9 44	7 52	4 9	27
4	0 17	2 17	3 43	4 10	3 30	1 52	0 43	5 49	9 5	9 43	7 46	4 0	26
5	0 21	2 21	3 45	4 10	3 27	1 48	0 55	5 57	9 8	9 42	7 40	3 51	25
6	0 26	2 25	3 47	4 9	3 25	1 44	1 6	6 6	9 12	9 40	7 34	3 43	24
7	0 30	2 28	3 48	4 9	3 22	1 40	1 17	6 14	9 16	9 38	7 27	3 34	23
8	0 34	2 31	3 50	4 9	3 20	1 36	1 28	6 23	9 19	9 36	7 21	3 25	22
9	0 38	2 35	3 52	4 8	3 17	1 31	1 39	6 31	9 22	9 34	7 14	3 16	21
10	0 43	2 38	3 53	4 7	3 14	1 27	1 50	6 39	9 25	9 32	7 8	3 7	20
11	0 47	2 42	3 55	4 7	3 11	1 23	2 0	6 47	9 28	9 30	7 1	2 58	19
12	0 51	2 45	3 56	4 6	3 9	1 19	2 11	6 55	9 31	9 27	6 54	2 49	18
13	0 55	2 48	3 58	4 5	3 6	1 15	2 22	7 2	9 33	9 24	6 47	2 40	17
14	0 59	2 51	3 59	4 4	3 3	1 10	2 33	7 10	9 35	9 21	6 40	2 30	16
15	1 3	2 54	4 0	4 3	3 0	1 6	2 48	7 17	9 37	9 18	6 33	2 21	15
16	1 8	2 57	4 2	4 2	2 57	1 2	2 54	7 24	9 39	9 14	6 26	2 12	14
17	1 12	3 0	4 3	4 1	2 53	0 57	3 4	7 31	9 41	9 11	6 18	2 3	13
18	1 16	3 3	4 4	4 0	2 50	0 53	3 14	7 38	9 42	9 7	6 11	1 54	12
19	1 20	3 6	4 5	3 58	2 47	0 49	3 25	7 45	9 44	9 3	6 3	1 44	11
20	1 24	3 9	4 6	3 57	2 44	0 44	3 35	7 51	9 45	8 59	5 56	1 35	10
21	1 28	3 12	4 6	3 55	2 40	0 40	3 45	7 58	9 46	8 55	5 48	1 26	9
22	1 32	3 14	4 7	3 54	2 37	0 36	3 55	8 4	9 47	8 50	5 40	1 16	8
23	1 36	3 17	4 7	3 52	2 33	0 31	4 5	8 10	9 47	8 46	5 32	1 7	7
24	1 40	3 20	4 8	3 51	2 30	0 26	4 15	8 16	9 48	8 41	5 24	0 57	6
25	1 44	3 22	4 9	3 49	2 26	0 22	4 25	8 21	9 48	8 36	5 16	0 48	5
26	1 48	3 25	4 9	3 47	2 22	0 18	4 35	8 27	9 48	8 31	5 8	0 38	4
27	1 51	3 27	4 10	3 45	2 19	0 13	4 45	8 32	9 48	8 26	5 0	0 29	3
28	1 55	3 30	4 10	3 43	2 15	0 9	4 54	8 37	9 48	8 20	4 51	0 19	2
29	1 59	3 32	4 10	3 41	2 11	0 4	5 4	8 42	9 47	8 15	4 43	0 10	1
30	2 3	3 34	4 10	3 39	2 7	0 0	5 13	8 47	9 47	8 9	4 35	0 0	0
	s. +11	s. +10	s. +9	s. +8	s. +7	s. +6	s. —11	s. —10	s. —9	s. —8	s. —7	s. —6	

Elements of Eclipses.

Anom.	Sun's Hor. Mot.	Moon's Hor. Mot.	Sun's Semi-Dia.	Moon's Semi-Dia.	Hor. Par.	Dark Cone	Light Cone	Anom.
s. °	′ ″	′ ″	′ ″	′ ″	′ ″	′ ″	′ ″	s. °
0 0	2 23	29 35	15 45	14 42	53 53	38 2	69 44	12 0
5	2 23	29 35	15 46	14 42	53 53	38 3	69 44	25
10	2 23	29 37	15 46	14 43	53 56	38 5	69 45	20
15	2 23	29 41	15 46	14 44	53 59	38 8	69 50	15
20	2 23	29 47	15 46	14 45	54 4	38 13	69 55	10
25	2 23	29 53	15 47	14 47	54 10	38 19	70 1	5
1 0	2 24	30 2	15 48	14 49	54 18	38 27	70 9	11 0
5	2 24	30 12	15 48	14 51	54 27	38 36	70 18	25
10	2 24	30 24	15 49	14 54	54 37	38 46	70 28	20
15	2 24	30 36	15 50	14 57	54 49	38 58	70 40	15
20	2 25	30 50	15 51	15 1	55 1	39 10	70 52	10
25	2 25	31 6	15 52	15 5	55 15	39 24	71 6	5
2 0	2 25	31 22	15 53	15 9	55 30	39 39	71 21	10 0
5	2 26	31 40	15 54	15 13	55 46	39 55	71 37	25
10	2 26	31 58	15 56	15 17	56 2	40 11	71 53	20
15	2 26	32 19	15 57	15 23	56 20	40 29	72 11	15
20	2 27	32 39	15 58	15 27	56 38	40 47	72 29	10
25	2 27	33 0	16 0	15 33	56 58	41 7	72 49	5
3 0	2 28	33 23	16 1	15 38	57 17	41 26	73 8	9 0
5	2 28	34 44	16 3	15 43	57 36	41 45	73 27	25
10	2 29	34 7	16 4	15 49	57 56	42 5	73 47	20
15	2 29	34 30	16 5	15 54	58 16	42 25	74 7	15
20	2 29	34 53	16 7	15 59	58 36	43 45	74 27	10
25	2 30	35 15	16 8	16 4	58 56	43 5	74 47	5
4 0	2 30	35 38	16 9	16 9	59 13	43 22	75 5	8 0
5	2 31	35 59	16 11	16 15	59 32	43 41	75 23	25
10	2 31	36 19	16 12	16 20	59 50	43 59	75 41	20
15	2 31	36 39	16 13	16 24	60 6	44 15	75 57	15
20	2 32	36 56	16 14	16 28	60 21	44 30	76 12	10
25	2 32	37 13	16 15	16 32	60 35	44 44	76 26	5
5 0	2 32	37 28	16 15	16 35	60 46	44 55	76 38	7 0
5	2 32	37 40	16 16	16 38	60 57	45 6	76 48	25
10	2 33	37 51	16 17	16 40	61 6	45 15	76 56	20
15	2 33	38 0	16 17	16 42	61 13	45 22	77 4	15
20	2 33	38 6	16 18	16 42	61 18	45 27	77 9	10
25	2 33	38 9	16 18	16 45	61 21	45 30	77 12	5
6 0	2 33	38 12	16 18	16 45	61 23	45 32	77 14	6 0

John Taylor, Printer, 24, Gray's Inn Road, London, W.C.

Breinigsville, PA USA
14 June 2010
239860BV00001B/9/A